# SpringerBriefs in Finance

More information about this series at http://www.springer.com/series/10282

Angelo Corelli

# Inside Company Valuation

Springer

Angelo Corelli
Department of Finance and Accounting,
School of Business Administration
American University in Dubai
Dubai
UAE

ISSN 2193-1720 ISSN 2193-1739 (electronic)
SpringerBriefs in Finance
ISBN 978-3-319-53782-5 ISBN 978-3-319-53783-2 (eBook)
DOI 10.1007/978-3-319-53783-2

Library of Congress Control Number: 2017933438



Printed on acid-free paper

This Springer imprint is published by Springer Nature
The registered company is Springer International Publishing AG
The registered company address is: Gewerbestrasse 11, 6330 Cham, Switzerland

*An investment in knowledge pays the best interest*

B. Franklin

# Preface

## A Modern Approach

The book offers an innovative view of the area of company valuation for newcomers in finance. The book aims to be a guide through the most common and interesting methods for corporate valuation, supported by a rich mathematical foundation.

The focus is on the tools and methods for valuing a business, from the most basic approaches to the implementation of complex algorithms. Each topic is deeply analyzed to give students the tools to really understand the effects of valuation on the life of the corporation.

The vision of the book is to offer a guide to valuation, capable of helping students and practitioners gain an introduction to the area, as well as giving intermediate users useful insights into and a deeper understanding of the analytics behind each part.

The literature of books in corporate valuation spans over a wide range of methods and theories. The market is already full of (even good) textbooks facing the issue of the corporation from many points of view.

However, the overall offer of *Inside Company Valuation* turns the attention on the financial aspects of the daily life of the corporation, with a robust mathematical setting, and the explanation and derivation of the mostly popular models of the firm.

The potential of filling such a niche is evident if one thinks that a course in Corporate Finance nowadays makes only sense if it includes a robust section about company valuation tools and techniques.

There is a demand of high-quality education at Master level, including a more advanced level of Corporate Finance courses, which is not fulfilled nowadays, and definitely requires a good textbook as a guide for such advanced education in the field.

Corporate finance is the first area of knowledge in an ideal sequence of studies in finance. Company valuation is a major component of the theory behind the broadest area and it is therefore crucial for all students to gain command of the topic.

The general purpose of a good textbook is to include all necessary knowledge without missing any crucial information, but also avoiding useless widening of the explanation beyond what can really satisfy the needs of the audience.

The structure of the book corresponds to the above rationale, giving a complete view of the various topics that make up the valuation task, and offering a good quantitative basis, which helps students understand the topics.

## An Innovative Pedagogy

The book stands on a double pillar of theory and analytics, which merge in a way that makes it easy for students to understand the exact meaning of the concepts and their representation and applicability in real-world contexts. Examples are given throughout the chapters in order to clarify the most intricate aspects; where needed, there are appendices at the end of chapters that give more mathematical insights about specific topics.

The duality stands at the basis of a complete learning experience, and students may get the most out of the book if they have a prior solid background in economic theory and an introductory level of financial mathematics. Indeed, students facing financial topics for the first time may benefit from using the book as a medium-level introduction to some aspects of financial theory and practice.

Due to the recent growth in knowledge demand by the private sector, practitioners can also benefit from the chapters in the book, and fill a gap between university and industry, which stood prior to the crisis. The book provides useful information for managers who want to increase their knowledge about risk management and understand what may have been lacking in their own systems.

## A Selected Audience

The book is meant for third-year undergraduate students of business finance, quantitative finance, and financial mathematics, as well as first-year postgraduate (master) students. Most universities offer the type of training in mathematics and statistics that would be prerequisite for the successful completion of a course using *Inside Company Valuation*. Potential users include students of universities, technical schools, and business schools offering courses in financial risk management.

The book represents a unique and innovative approach to the field of corporate finance. Competitors, in fact, focus on the business side of the story, centering the discussion on corporate aspects only. There is no challenge or criticism of the markets side, and there is no drive to understand the mathematical foundations of theory in a critical sense. That is exactly what *Inside Company Valuation* wants to offer instead.

A quantitative approach incorporates a more critical view, contributing to a description of theory that does not blindly rely on numbers, indices, and ratios, but takes into account the variety of (sometimes unpredictable) situations that characterize financial markets.

Certainly, it is not the typical corporate finance-related brief, but it is a book that never gives up on the reader. Even in the most complicated parts, which are anyway at an intermediate level, students are guided through the processes and given the tools they need; nothing is cryptic.

Although the market for books in corporate finance is heavily crowded, *Inside Company Valuation* aims to distinguish itself through the higher level of mathematics involved, compared to the competition.

Moreover, the links to real industry examples are much more updated than other texts can offer at the moment. Therefore, a niche in the market can be envisioned for a book that challenges students on a more analytical than business-related level.

## A Reliable Partner for Instructors

*Inside Company Valuation* is mostly tailored for in-class lectures, yielding the best learning experience when combined with good quality lectures. The overall flexibility of the book, in five chapters, and the straightforward structure also make it a good reference for online learning. However, the medium-high level of difficulty of the book suggests the need for a closer relation with the instructor and the possibility of in-person explanations.

The structure of the book is such that a typical module of six ECTS and approximately 30 hours of front teaching would get partial coverage, for the part about company valuation. The five chapters (at most 15 pages each) fit in a course design in the range of about 2–3 lectures of 1.5 hours of effective teaching. The structure also fits the international standard of a course with two lectures per week spanned over a two-month teaching term. The overall contents of the book can fill approximately 4–6 hours of teaching.

Dubai, UAE Angelo Corelli

# Acknowledgements

First of all I would like to thank my family for being at my side all the time. My father **Giuseppe**, my mother **Rina**, and my brother **Mauro** always gave me all the support I needed to move forward in my career.

I would like to thank Dr. Lance **De Masi**, President of the American University in Dubai (AUD), UAE, for managing the university in a way that creates the perfect environment for developing this book.

A special thanks to Dr. Jihad **Nader**, Provost and CAO of AUD, for hiring me and giving the trust and support that help me working every day.

I want to thank Dr. Mohammed **Abu Ali**, Dean of the School of Business Administration at AUD, for his direct help and support in the development of my career path in the University. Also thanks to Dr. Assaad Farah for his continuous support and the openess of mind about Academic life and support to the faculty.

Also, a special thanks to Dr. Jatin **Malhotra** (Jay), Chair of the Finance and Accounting Department at the SBA, AUD, for the daily support and the practical hints for surviving my first months in a new environment.

I would also like to offer a special thanks to Dr. Ann Marie **Simmonds**, Assistant Professor of English, School of Arts and Sciences, AUD, for the precious help in proofreading this book.

A special thanks to all my **friends and relatives** who supported me during this journey, and to all my **students**, with whom I engaged in fruitful discussions related to the topics covered by this book.

And last but not least, my thoughts go to the team of Springer that brilliantly contributed to this project, including the Associate Editor Rocio **Torregrosa**, the Senior Editorial Assistant Irene **Barrios-Kezic**, and all the production team.

Angelo Corelli

# Contents


**1 The Value of the Firm** .......... 1
1.1 The Determinants of Corporate Value .......... 2
1.2 Balance-Sheet-Based Methods .......... 4
1.3 Income-Statement-Based Methods .......... 6
1.4 Goodwill-Based Methods .......... 8
Bibliography .......... 12

**2 Dividend-Based Valuation** .......... 15
2.1 Dividend Discount Models .......... 16
2.2 The Gordon Growth Model .......... 18
2.3 Multiple-Stage Growth Model .......... 21
2.4 Dividend Valuation Models and Market Efficiency .......... 25
Bibliography .......... 27

**3 Free-Cash-Flow-Based Methods** .......... 29
3.1 The Determinants of Cash Flows .......... 30
3.2 The Growth Rate .......... 33
3.3 Cash Flows and Corporate Value .......... 37
Bibliography .......... 39

**4 The Valuation of Private Firms** .......... 41
4.1 Beta Estimation .......... 42
4.2 Cost of Debt Estimation .......... 48
4.3 Growth and Terminal Value .......... 50
Bibliography .......... 58

**5 A Real Case: Facebook** .......... 59
5.1 A Strange IPO .......... 60
5.2 The Firm's Fundamentals .......... 63
5.3 The Solution to the Case Study .......... 64
Bibliography .......... 68

# Chapter 1
# The Value of the Firm

**Abstract** There are several factors that determine the value of a company, all important and relevant in order to achieve the full exploitation of corporate assets. Some of them are the history and nature of the business, the economics of the industry, the book value and financial condition of the business. Valuation can be based on several methods, and it is possible to perform corporate valuation starting from either the balance sheet of the company, or its income statement. Another class of methods involves the goodwill as a measure of the projected corporate profitability and therefore, as the primary figure that adds up to the discounted cash flows from income. Overall, all the presented methods rely on the knowledge of the financial situation of the company and the ability to project the figures to the immediate future in order to define scenarios for the future profitability of the firm.

**Keywords** Value · Worth · Wealth · Balance sheet · Price · Drivers · Income statement · Goodwill

A big part of corporate finance knowledge focuses on corporate valuation, which represents one of the most challenging tasks to perform in relation to the analysis of the company financials.

The methods for valuating the firm rely on a mix of art and science, with several input variables involved. The rest is about intuition and correct understanding of the parameters estimation.

Many companies do professional valuation for acquisitions, mergers and IPOs, but very rarely different valuations on the same company reach a consensus about the corporate value.

The process of valuating a firm goes through several stages, and implies good estimation of several key parameters, like the growth potential of the firm, the size of the cash flows associated to the business, the right discounting factors and more.

The sense of going through such a process is to achieve an educated guess on what the theoretical value of a company should be based on the available information, the balance sheet, and the robust theoretical background developed in recent years.

 
A. Corelli, *Inside Company Valuation*, SpringerBriefs in Finance,
DOI 10.1007/978-3-319-53783-2_1

Good financial management is about knowing the value of the assets and the equity of the company that is under management, therefore requiring the knowledge of the appropriate methods for understanding those values.

After studying this chapter, you will be able to answer the following questions, among others:

1. What are the factors that determine the corporate value?
2. What are the methods for valuation based on the balance sheet?
3. What are the methods for valuation based on the income statement?
4. What are the methods for valuation based on the goodwill?
5. How can the different models implemented and integrated to determine the fair value of the company?

The first section of the chapter is about the methods for company's valuation based on the balance sheet. The second section is about the methods for company's valuation based on the income statement. The third section focuses on the methods for company's valuation based on the corporate goodwill.

## 1.1 The Determinants of Corporate Value

From an accounting point of view the value of the business for the shareholders is given by the difference between the corporate assets and corporate liabilities. But this is not exactly what the business is worth, in all cases, for several reasons.

Another way to value the business in fact is its selling price, in case a sale takes place. At that time, one will be sure about the value of the company, given that there is some other agent willing to pay that specific price for it.

But the best way to assess the value of a company is probably to perform a valuation. The scope, detail and cost of a business valuation can vary dramatically based on the size and complexity of the business being considered, on the purpose of the valuation.

It is important to understand some terminology issues, especially regarding the difference between words like "value" and "worth", which in standard conversation are normally interchangeable.

However, when it comes to economics framework and valuation issues, these terms end up having slightly different meanings. The differences can be explained by having a closer look at the financial and economic meaning of the words.

Fair market value is an hypothetical value that is based on modelling and represents a target or benchmark. Its calculation is subject to differing opinions and judgments—different professional valuations may yield different "fair market values."

There is normally no actual sale or transaction of any type supporting such a value, which remains ideal, as a judgment of the value of a business given by some analyst due to some specific model.

The worth is instead a real amount that can be confirmed by the result of some actual transaction, and matches the equilibrium between the asked and offered price from the parties involved in the deal.

The price is the effect of the worth in terms of the actual number assigned to the real value. The asking price from a seller is the starting point for a negotiation, and the buyer will try to relaunch, as the seller will do as well as a follow up.

The result is a selling price where the buyer and seller end up, based on their individual judgments of worth and their skills as negotiators, in combination with the particular pressures that may be pushing them toward buying and selling.

Company valuation is not only useful to give an entrepreneur an idea of the value of the business. In some cases showing a good and fair corporate value allows companies to convince banks to issue a loan, based on creditworthiness.

In case an IPO is set, or more generally, the sale of a business is taking place, the valuation helps to set the range of values to be included in the negotiation process.

It can be a vital aid in the decision-making involved with putting a business on the market, but it doesn't replace negotiation.

The value as calculated through a valuation analysis also provides the liquidation value of a departing partner in the corporation. It helps to avoid costly litigation in case of a dispute, and it needs to be updated periodically.

Employees with stock ownership plans find valuation very useful to assess the value of the securities contributed by the company to the plan. It also gives a good insight of the value subject to taxation for particular assets.

In case the business capitalization, valuation provides the basis to asses the value of the stocks for different classes. It also establishes a value for the stock help by dissenting shareholders who may need to be bought out and it establishes a value for purposes of litigation.

The value is characterized by several components, which are listed by the International Reporting Standards (IRS) as:

- The history and nature of the business.
- The economics of the industry.
- Book value and financial condition of the business.
- Earnings and dividend paying capacities of the business.
- Goodwill and other intangibles.
- Sales of stock and size of block to be valued.
- Market price of actively traded stocks of public companies engaged in the same or a similar business.

In the IRS there is no quantitative modelling provided, but mostly a series of qualitative recommendations that can be summarized as "A sound valuation will be based upon all relevant facts … common sense, informed judgment, and reasonableness."

Actually, business valuation has its foundation in the capitalistic principle that investors invest in businesses for one reason: to generate wealth, usually measured in terms of profit.

Furthermore, expectations of profit vary based on the degree of investment risk—the higher the risk, the higher the expected profit demanded. In the next articles we'll take a look at various methods for determining the answer to that key question.

## 1.2 Balance-Sheet-Based Methods

In general there is no consensus about the right valuation of a firm. In most cases different economic agents come with different valuations, and in the case of acquisition or mergers, it is very typical that the buyer and seller come with different numbers.

Value should not be confused with price, which is the quantity agreed between the seller and the buyer in the sale of a company. Consider the case when a foreign company wishes to enter a new market, through buying a well-known local company in the target market.

The valuation of the local firm by the foreign buyer will be mostly focused on the brand value, but will not consider the plant, machinery and other assets, given the fact, that the buyer itself owns the assets. The seller point of view instead will be claiming the overall value of the firm as a sum of the goodwill and the tangible assets.

From the buyer's viewpoint, the basic aim is to determine the maximum value it should be prepared to pay for what the company it wishes to buy is able to contribute. There are therefore two extremes on the table of negotiations and normally the final price the parties agree on is somewhere between the two figures.

In financial portfolio management and trading, the valuation of a company's stock is normally used to asses whether the stock is currently overpriced or underpriced on the market, and take a position (long or short) on the stock according to the comparison.

By valuing different companies it is also possible to determine which securities should be included in the portfolio to invest in, especially considering that it can be remunerative to take long position on undervalued stocks.

The valuation of several companies is also used to make comparisons between companies. For example, if an investor thinks that the future course of Samsung's share price will be better than that of Apple, he may buy Samsung shares and short-sell Apple shares.

Valuation is also a tool to justify the price at which stocks are offered to the public, and it is fundamental for other tasks like separating the equity value from that of other assets and liabilities.

On top of that the valuation as a central role in quantifying the managerial contribution to the value creation in the firm, and determine the performance-related compensation owed to the managers.

From an analytical point of view the valuation of a company is an important tool to identify the value drivers. Moreover it gives a clear idea about whether to continue the business or not, grow or merge, and more.

The valuation of the company and the different business units is fundamental for deciding what products/business lines/countries/customers (and more) to maintain, grow or abandon. It also provides a means for measuring the impact of the company's possible policies and strategies on value creation and destruction.

There are several methods available for valuing the company's assets or equity. The first category resembles the balance-sheet-based methods, that are mostly used to value equity.

According to these traditional methods the value of the firm can be derived from the information in the balance sheet. The value is determined from a static viewpoint, which, therefore, does not take into account the company's possible future evolution or money's temporary value.

Other factors like macroeconomic drivers, the industry's current situation, the state of human resources and the managerial issues are also ignored, in that they are not listed in any financial statement.

Some of the balance-sheet-based methods are the following: book value, adjusted book value, liquidation value, and substantial value. The purpose of the following pages is to analyse each of them.

The book value of the company can be inferred from the value of the shareholder's equity stated in the balance sheet, namely the sum of capital and reserves. The value can also be defined as the difference between the firm's total assets and the firm's total liabilities.

*Example 1.1* The balance sheet of company AZC shows figures as per the Table 1.1.

The book value of the corporate equity is €160M, and it is also the difference between total assets (€320M) and total liabilities (€160M). This value suffers from the shortcoming of its own definition criterion: accounting values are different from the market values, with the result that the book value almost never matches the "market" value.

In order to overcome the downside of pure book value approach, the method called Adjusted Book Value can help. When the values of assets and liabilities match their market value, the adjusted net worth is obtained.

**Table 1.1** An example of corporate balance sheet

| Assets (€M) | | Liabilities and equity (€M) | |
|---|---|---|---|
| Cash | 10 | Accounts payable | 80 |
| Accounts receivable | 20 | Bank debt | 20 |
| Inventories | 90 | Long-term debt | 60 |
| Fixed assets | 200 | Equity | 160 |
| Total assets | 320 | Total liabilities and equity | 320 |

**Table 1.2** An example of corporate balance sheet after adjusting for market values

| Assets (€M) | | Liabilities and equity (€M) | |
|---|---|---|---|
| Cash | 10 | Accounts payable | 80 |
| Accounts receivable | 16 | Bank debt | 20 |
| Inventories | 104 | Long-term debt | 60 |
| Fixed assets | 300 | Equity | 270 |
| Total assets | 430 | Total liabilities and equity | 430 |

*Example 1.2* Starting from the numbers in Example 1.1, assume that company AZC has €4M of bad debt, that reduce the value of debt. Also consider that the inventory must be revalued accounting for a total value of €104M. Moreover, assume an expert has given an estimate of the effective value of corporate assets, of €300M. Given the above adjustments the new balance sheet is shown in the Table 1.2.

The liquidation value of a company, as the name suggests, is the result of the valuation done at the time of dismissal of a business. It is calculated by deducting the liquidation expenses and all the payments due to the various stakeholders, from the overall value of the company at the time of liquidation.

*Example 1.3* Taking the example given in Table 1.2, if the redundancy payments and other expenses associated with the liquidation of the company AZC were to amount to $120M, the shares' liquidation value would be €150M (270–120).

Such a method is of course only useful in case of corporate liquidation, and does not apply to any other situation. However, it always represents the company's minimum value as a company's value, assuming it continues to operate, is greater than its liquidation value.

The substantial value represents the amount of investment that should be put in place by an entrepreneur that aims at forming a company with identical features to the firm that is being valued.

Compared to the liquidation value, the substantial value, also called assets' replacement value is based on the assumption that the firm continues running the business. Such a value does not normally include things like unused assets. The Gross substantial value is the assets' value at market price while the net substantial value is the gross substantial value minus the value of liabilities.

## 1.3 Income-Statement-Based Methods

Another class of valuation methods is based on the Income Statement. As opposed to the Balance Sheet methods, the Income Statement methods rely on the information inferred from the income statement.

**Table 1.3** An example of corporate income statement

| Income statement (€M) | |
|---|---|
| Sales | 150 |
| Cost of sales | (56) |
| General expenses | (50) |
| *EBIT* | *44* |
| Interest expenses | (4) |
| *EBT* | *40* |
| Taxes (35%) | (14) |
| *Net Income* | *26* |

They try to determine the company's value through the earnings, sales and other figures. As an example, a common practice in company valuation is to perform quick valuations.

Another quick valuation can be done by valuing the number of service slots available to a service company by a multiple. Also, insurance companies can be valued by multiplying annual premiums by a multiple.

*Example 1.4* The income statement for company AZC is given in the Table 1.3.

The Price-Earnings Ratio (PER) is one of the possible methods based on income statement. It multiplies the net income by the PER in order to obtain the value of the equity as from

$$E = PER \times X$$

where

$E$ is the value of the equity
$PER$ is the price-earnings ratio
$X$ is the value of the earnings

The PER approach is one of the simplest methods based on earnings and it only gives a rough estimation of the equity value of the firm. However its simplicity has made it a popular tool for quick preliminary valuation.

Dividends are another important variable of interest in the consideration of the net income made by the business. They represent the part of income that is paid out to shareholders.

Dividends are normally not paid regularly but the valuation methods that are based on dividends rely on the assumption that a constant dividend is paid to the shareholders along the duration of the investment.

According to the dividend-based valuation method, a share's value is the net present value of the dividends that we expect to obtain from it. The valuation of the cash flows that dividends represent can be done by using simple time value calculations like annuity and perpetuity.

In the perpetuity case for example, that is, a company from which we expect constant dividends every year, the value can be expressed as follows:

$$E = \frac{D_0}{r_E}$$

where

$D_0$ is the value of the distributed dividends
$r_E$ is cost of equity capital (required return on equity)

By recalling the formulas for a growing perpetuity we can introduce the additional assumption that the dividends paid to the shareholders will grow at some specified growth rate. The formula for equity valuation then becomes

$$E = \frac{D_1}{r_E - g}$$

where

$D_1$ is the value of the projected future dividends
$g$ is the constant growth rate of dividends in the future

Some companies, from some industry sectors, use the Sales Multiples method for valuation. It consists of calculating the value of a company as the product of its sales by some fixed number.

The fixed number normally changes from one industry sector to the others, and for example it is equal to 2 for pharmacies under some market conditions. But it varies for other sectors. It is also a common practice to value a soft drink bottling plant by multiplying its annual sales in liters by 500 or another number, depending on the market situation.

Valuation through multiples can be also done by using other variables like the PER, the Price-to-Sales Ratio (PSR), Value-to-EBIT Ratio (VER) and more. The obvious caveat is that in order to carry on multiples-based valuation of some companies, the multiples of comparable companies must be used.

## 1.4 Goodwill-Based Methods

Goodwill is the added value on top of the book value, meaning that part of intangible assets that does not go into the balance sheet but it is represented by reputation and other things.

Goodwill seeks to represent the value of the company's intangible assets, which contribute an advantage with respect to other companies operating in the industry. Goodwill items include (but are not limited to) quality of the customer portfolio, industry leadership, brands and strategic alliances.

Determining the value of the goodwill can be an extremely hard task, and there is no consensus among the right choice of model to value it. Some of the methods used to value the goodwill give rise to the various valuation procedures described in this section.

Methods embracing the inclusion of the goodwill in the corporate value adopt a mixed approach to valuation. On the one hand they give a static measurement of the value of the company's assets. On the other hand they project the future revenues of the company to complete the task.

Basically, these methods seek to determine the company's value by estimating the combined value of its assets plus a capital gain resulting from the value of its future earnings: they start by valuing the company's assets and then add a quantity related with future earnings.

According to the Classic Valuation Method the value of a firm is given by the sum of the net assets and the goodwill. The value of the latter is obtained as a multiple of the company's net income. The formula that expresses a company's value is:

$$V = E + (k \times X)$$

where

$E$ is the value of the corporate net assets
$k$ is a coefficient between 1.5 and 3.0

The formula is mainly used for industrial companies, while it can also be used for the retail trade by expressing it in another form, as

$$V = A + (s \times Z)$$

where

$s$ is the percentage of sales revenue
$Z$ is the company's turnover

*Example 1.5* According to the Classic valuation method, assuming the goodwill to be equal to three times the net income, the value of company AZC is equal to

$$V = 270{,}000{,}000 + (3 \times 26{,}000{,}000) = \text{EUR}348{,}000{,}000$$

A modified version of the goodwill model, called Abbreviated Goodwill Income (AGI) method, is based on another formula, namely

$$V = E + PV_A \times (X_{-1} - r_{ALT} \times E)$$

where

$PV_A$ is the present value of an annuity between 5 and 8 years
$X_{-1}$ is the net income for the previous year
$r_{ALT}$ is the interest rate available on alternative investments

Clearly the product $PV_A \times (X_{-1} - r_{ALT} \times \bar{A})$ represents the goodwill of the company. A possible interpretation of the formula is that the value of the firm is given by the sum of its adjusted net worth and the goodwill. The value of the goodwill is given by some coefficient applied to the difference between net income and investment in the net assets *A*. The investment bears an interest *i* corresponding to the risk-free rate.

*Example 1.6* According to the AGI valuation method, assuming that the annuity is on 5 years at 15% interest rate, the following holds

$$PV_A = 3.352$$

and if the interest rate $r_{ALT}$ is equal to 10%, the equity's value of company AZC is equal to

$$\begin{aligned} V &= 270{,}000{,}000 + 3.352 \times (26{,}000{,}000 - 0.10 \times 270{,}000{,}000) \\ &= 270{,}000{,}000 - 3{,}352{,}000 = €266{,}648{,}000 \end{aligned}$$

Another method is based on the approach recommended by the Union of European Accounting Experts (UEC). According to the UEC the company's value is obtained from the equation

$$V = \left[\frac{E + (PV_A \times X_{-1})}{1 + (PV_A \times r_{ALT})}\right]$$

The formula can be interpreted as the sum of the substantial value of the firm, calculated as the revalued assets, and the goodwill. The latter is calculated by capitalizing at compound interest (using the factor $PV_A$) the difference between the profit obtained by investing at a risk-free rate *i* a capital equal to the company's value *V*.

*Example 1.7* According to the UEC method, the value of the equity of company AZC would be

$$V = \frac{270{,}000{,}000 + (3.352 \times 26{,}000{,}000)}{1 + (0.10 \times 3.352)} = \frac{357{,}152{,}000}{1.3352} = €267{,}489{,}515$$

The Indirect Method (IM) for valuing the company can be expressed by the formula

$$V = E + \left( \frac{X_{-1} - (r_{ALT} \times E)}{2r_{ALT}} \right)$$

In this case the $i$ is the interest paid on long term Treasury bonds.

The method implies that the value of the net assets and the value of the return are equally weighted, and it is very flexible in that it allows to vary the weights in order to obtain other different variants.

*Example 1.8* According to the Indirect method, the value of the equity of company AZC would be

$$\begin{aligned} V &= 270{,}000{,}000 + \frac{26{,}000{,}000 - (0.10 \times 270{,}000{,}000)}{2 \times 0.10} \\ &= 270{,}000{,}000 - \frac{1{,}000{,}000}{0.20} = €265{,}000{,}000 \end{aligned}$$

The Direct Method (DM) for company valuation is described by the formula

$$V = E + \left( \frac{X_{-1} - (r_{ALT} \times E)}{\alpha \times r_{ALT}} \right)$$

where

$\alpha$ is a risk-weighting coefficient between 1.25 and 1.50

*Example 1.9* According to the DM, the value of the equity of company AZC, assuming a risk coefficient equal to 1.50, is given by

$$\begin{aligned} V &= 270{,}000{,}000 + \frac{26{,}000{,}000 - (0.10 \times 270{,}000{,}000)}{1.5 \times 0.10} \\ &= 270{,}000{,}000 - \frac{1{,}000{,}000}{0.15} = €263{,}333{,}334 \end{aligned}$$

The goodwill is here expressed by restating the difference between net income and the potential profit obtained if investing the corporate assets at an interest $i$, for an indefinite period of time.

With the Annual Profit Purchase (APP) method the formula for the valuation is

$$V = E + n \times [X_{-1} - (r_{ALT} \times E)]$$

where

$n$ is the number of years the super-profit is realized

This method implies that the value of the goodwill is equal to the sum of the superprofits over some time interval. The value to be paid by the buyer is then obtained as the sum of the value of net assets plus the total amount of superprofits.

The superprofit is calculated on a number of years normally ranging between 3 and 5, and the interest rate for long-term borrowing applies.

*Example 1.10* According to the APP method, the value of the equity of company AZC, assuming superprofits are consistent for 5 years, is given by

$$\begin{aligned} V &= 270{,}000{,}000 + 5 \times [26{,}000{,}000 - (0.10 \times 270{,}000{,}000)] \\ &= 270{,}000{,}000 - 5{,}000{,}000 = €265{,}000{,}000 \end{aligned}$$

The final class of methods presented in this section is the family of the Cash Flow Discounting-based (CFD) methods. The valuation in this case is based on the present value of the future cash flows generated by the firm. A complete analysis of the method is given in Chap. 3. The purpose of this section is to introduce the methodology.

The methods introduced before are old fashioned and have been used in the past decades. Nowadays most of the valuation related to mergers, acquisitions and IPOs are carried through CFD modelling.

In these methods, the company is viewed as a cash flow generator and the company's value is obtained by calculating these flows' present value using a suitable discount rate.

For each time spot included in the analysis period, cash flow discounting methods are based on the forecast of the financial figures related to corporate operations, including sales, cost of workforce, inventory, administration, loan repayments etc.

In cash flow discounting-based valuations, a suitable discount rate is determined for each type of cash flow. Determining the discount rate is one of the most important tasks and takes into account the risk, historic volatilities; in practice, the minimum discount rate is often set by the interested parties (the buyers or sellers are not prepared to invest or sell for less than a certain return, etc.).

## Bibliography

Brealey RA, Myers SC (2000) Principles of corporate finance, 6th edn. McGraw-Hill, New York

Copeland TE, Koller T, Murrin J (2000) Valuation: measuring and managing the value of companies, 3rd edn. Wiley, New York

Damodaran A (2002) Investment valuation, tools and techniques for determining the value of any asset, 2nd edn. Wiley, New York, p 300

Fernández P (2004) Valuing companies by cash flow discounting: ten methods and nine theories. SSRN Working Paper no. 256987

Fernández P (2007) Company valuation methods. The most common errors in valuation. IESE BS WP No 449

Pratt SP (2008) Valuing a business: the analysis and appraisal of closely held companies, 5th edn. McGraw-Hill, New York, p 243

# Chapter 2
# Dividend-Based Valuation

**Abstract** Dividend discount models for equity valuation are a popular tool in the analysis of corporations and their financials. By using dividends as the target cash flow for the calculation of the present values, one directly aims at the equity value. The simplest forms of DDM rely on the discount of a perpetuity, by assuming the company will pay a constant dividend overtime, during its life. When the perpetuity allows for a constant growth rate of dividends overtime, it translate into the Gordon Growth model, which represents the value of the corporate equity as the present value of a growing perpetuity. Most of the multi-stage DDM are in fact designed to allow for an initial period of discrete cash flows to be discounted at the appropriate rate, and be summed up with the present value of a final perpetuity calculated on the residual dividend after the discrete period.

**Keywords** Dividends · Permanent growth · Perpetuity · Premium · Risk-free rate · Multiple stages · Efficiency

Dividends are the most straightforward type of cash flows produced by a company as a consequence of profitable investments. It therefore comes natural to adopt valuation models based on dividends.

The Dividend Discount Models are a powerful tool to discern the intrinsic value of the company based on the corporate payout policy, and the size and structure of the payments.

Either based on constant or growing dividends, all the models presented in this chapter share the practicality of application and the accessible level of the math required to perform them.

Depending on the number of stages considered for the model, the formulas change but the sense of the analysis does not. Dividends still represent a good base to start from for corporate valuation.

After studying this chapter, you will be able to answer the following questions, among others:

1. How can dividends be used for value calculation?
2. How can the model parameters be estimated?



A. Corelli, *Inside Company Valuation*, SpringerBriefs in Finance,
DOI 10.1007/978-3-319-53783-2_2

3. What is the role of the growth of dividends over time?
4. What are the main drivers of the dividend payout decision and how does it relate to value calculation?
5. What are the different models available for dividend-discount valuation?

The first section of the chapter introduces the dividend-based discount modelling of corporate value. The second section is about the Gordon growth dividend discount model and its application to growing dividends overtime. The third section focuses on multiple-stage models where a discrete initial growth period is combined with an infinite growing period in perpetuity. The final section deals with considerations about the use of dividend discount models and further issues related to market efficiency.

## 2.1 Dividend Discount Models

All Dividend Discount Models (DDMs) are based on the discounting of a stream of future dividend payments, in order to calculate a present value that represents the value of the corporate equity.

Another important variable in the DDM models is the return on equity (cost of equity), the demanded return from an investor seeking to invest in the company. It is the compensation that the investor claims in exchange of the capital inflow.

It is therefore very important to address the cost of equity and be able to calculate it. According to the Capital Asset Pricing Model, the return on equity can be calculated by using the formula

Recall that the CML describes the relationship between the excess return of an asset and the excess return of the market, as a line in the risk-return space.

$$r_i - r_f = \beta\left(r_m - r_f\right)$$

where

$r_m$ is the return of the market.
$\beta$ is the dependence factor between market and asset.

Once the cost of equity is calculated it can be used to discount the cash flows represented by the dividends. A popular method in the class of the DDMs is the dividend growth model (DGM), that measures the stock price and capital to be raised by using the information about the dividends paid out to shareholders by the company.

$$P_0 = \frac{D_0(1+g)}{r_e - g} \tag{2.1}$$

where

$D_0$ is the current dividend
$g$ is the expected dividend future growth rate
$r_e$ is the cost of equity.

As can be observed, the equation aims to predict the ex-dividend market price of the share by processing the dividend today, the expected future dividend growth and the appropriate cost of equity.

The numerator indicates the expected dividend in one period, due to the growth $g$. It is possible to rework Eq. (2.1) in order to isolate the cost of equity. When the current price of the equity (stock) is known, the model can then be used to estimate the cost of equity, and this model can take into account the dividend growth rate.

$$r_e = \frac{D_0(1+g)}{P_0} + g$$

Since all the terms on the right hand side are known at time zero, for a listed company, the calculation of the cost of equity is straightforward. When other information is not available, the future dividend growth is assumed to continue at the recent historical rate.

*Example 2.1* The management of company ABC decides to pay a dividend of €1.75 on a stock with market price equal to €23.75 (cum-dividend). The historical dividend growth rate, which is expected to continue in the future, is 2.25%. Using the information it's possible to calculate the return on equity as

$$r_e = \frac{1.75(1+0.0225)}{23.75} + 0.0225 = 0.0978 = 9.78\%$$

One way to make the calculation more accurate is to use the ex-dividend price of the stock, obtained by subtracting the coming dividend from the cum-dividend price of the stock.

*Example 2.2* Following the previous example, the ex-dividend price is given by

$$P_{0,Ex} = 23.75 - 1.75 = €22.00$$

The cost of equity is, therefore given by:

$$r_e = \frac{1.75(1+0.0225)}{22.00} + 0.0225 = 0.1038 = 10.38\%$$

Stock valuation is based on the general assumption that if a market is populated by rational investors, the value of the stock today represents the present value of all future cash flows to be accrued in the future.

The rules of time travel allow to discount back all dividends in time and sum them up to a present value, so obtaining the price today of the stock. This is the intrinsic value of the stock, calculated on the basis of all available information.

The expectation of the investor who takes a long position on a stock is usually to get two types of cash flow in return. The first type is an income cash flow represented by the dividends paid out during the time span of the investment.

On top of that the investor can realize a capital gain by selling the stock back to the market when the price is convenient for that, and make a profit on the price difference compared to the buying price.

Since this expected price is itself determined by future dividends, the value of a stock is the present value of dividends through infinity.

$$P_0 = \sum_{t=1}^{\infty} \frac{D_t}{(1 + r_e)^t}$$

where,

$D_t$ is the expected dividends per share
$r_e$ is the cost of equity

Expected dividends and the return on equity are the two main components of the model, where the former are obtained through making assumptions about the expected future growth rates in earning and payout ratios.

About the cost of equity, it depends on how the asset is risky and it is normally measured with pricing model like the Capital Asset Pricing Model (CAPM), or other types of linear multivariate models.

The model is flexible enough to allow for time-varying discount rates, where the time variation is caused by expected changes in interest rates or risk across time.

The issue with projecting future dividends is complicated because it is not possible to make such a projection through infinity. Therefore, several versions of the dividend discount model are available, making different assumptions about future growth.

## 2.2 The Gordon Growth Model

The Gordon Growth Dividend Model (GGDM) is a model designed to be very simple and immediate. It values stocks in the framework of stable growth, for a company paying as much as it can in dividends.

The model can be used to value a firm that is in ‘steady state’ with dividends growing at a rate that can be sustained forever, and relates the value of a stock to its expected dividends in the next time period, the cost of equity and the expected growth rate in dividends.

$$P_0 = \frac{D_1}{r_e - g}$$

where,

$D_1$ is the dividend expected in the next period.

The meaning of a stable growth rate is a rate that spans on the company's dividends forever. The two insights in estimating such a rate also include the fact that all other financial figures are expected to grow at the same rate.

It is in fact the case that, when the dividends grow at a higher rate than the whole company and its earnings, the dividends will exceed the earnings over time. If the opposite holds, and the earnings grow at a higher rate than dividends, the latter will then converge towards zero, which is different from a steady state. In order for a company to grow at a steady state, the expected growth rate should be substituted in, so as to get exactly the same result for dividends growth and firm's growth.

There is also an issue related to how to judge whether a growth rate can be considered reasonable for a stable growth. The rule of thumb is that the growth rate has to be less than or equal to the growth rate of the economy in which the firm operates.

The agreement among analysts is not unanimous and the rate to be adopted in this case is controversial. Academicians however normally agree that there are three fundamental reasons for a company to be in a steady state.

Expected inflation is normally governed by a fair degree of uncertainty, making a difference in the real growth of the economy. With different expectations about the level of prices, there could be different projections about the growth rate of a country.

The growth rate of the company can be much less than the growth rate of the overall economy, and in some cases the firms become smaller over time, in proportion to the economy.

There is a final consideration regarding the fact that when a company is seen to be growing at a little bit more than a stable growth rate, for a few years continuously, then it is appropriate to assess the value of the firm by adding a premium to the stable growth rate. In the assumption of a non-stable growth of the company, then the analysis should be carried on in multiple stages.

In practice, no firm matches the assumption of constant growth rate, and earnings are normally quite volatile over the years. Anyway it is still possible to apply the model when the earnings show some constancy over cyclical periods.

Thus, a cyclical firm that can be expected to have year-to-year swings in growth rates, but has an average growth rate that which is defined, can be valued using the Gordon growth model, without a significant loss of generality.

A usual practice of companies is to smooth dividends over time. This is an argument in favour of the above theory and the fact that dividends are not linked to the cyclicality of earnings.

From the mathematical point of view, an average growth rate gives results that are not very different from ones obtained by using a constant growth rate. It is therefore possible to state the value of the stock as the present value of dividends per share paid in perpetuity, under the assumption that dividends are paid forever, as

$$P_0 = \frac{D_1^*}{r_e}$$

where

$D_1^*$ is the constant dividend per share expected from the next period.

The cost of equity that discounts the dividend in perpetuity is equivalent to the time value of the capital that is invested. It also reflects the risk of the investment.

The return and the maturity of the investment are directly related and the longer the money is tied, the higher the required return will be. The latter also is a consequence of the risk associated with the investment, due to the uncertainty about the future cash flows.

*Example 2.3* Assume a stock paying a current dividend of €1.75 per share and a required rate of return of 10.5%. The value of a share of stock is therefore

$$P_0 = \frac{1.75}{0.105} = €16.67$$

Therefore, by paying €16.67 per share and assuming constant future dividends of €1.75 per share, the investor will earn a 10.5% return per year on the investment every year.

If the assumption of a constant growth of the dividend over time is made, it is possible to factorize the amount of dividend $D_1$ at time 1 into the dividend at time zero multiplied by the growth factor.

In this case the value of the common stock at present is given by discounting an amount of infinite dividends growing at a constant rate $g$. The obvious way to perform such a calculation is through a growing perpetuity. The specific case of dividends growing at the constant rate $g$ is commonly referred to as the dividend valuation model (DVM):

$$P_0 = \frac{D_0(1+g)}{r_e - g} = \frac{D_1}{r_e - g}$$

This model is also referred to as the Gordon model and it is one of a general class of models referred to as the dividend discount model (DDM).

*Example 2.4* Astock pays a current dividend of €1.75 and the dividends per share are expected to grow at a rate of 2.25% per year. If the required rate of return (cost of equity) is 10.5% in perpetuity, the value of the share is

$$P_0 = \frac{1.75(1 + 0.0225)}{0.105 - 0.0225} = €21.69$$

If the growth is not expected to be constant, but rather growing over time, as it happens in real cases, the model changes. Companies normally experience seasonalities and cyclicalities that make the growth rate change over time.

Normally companies experience initial rapid growth when they start, a decreased growth in the intermediate phase of operations, and a situation of declining growth in their final stage. Further, companies may experience changes in their growth due to acquisitions and divestitures.

## 2.3 Multiple-Stage Growth Model

It is possible to modify the Dividend Growth model in order to allow for a two-stage representation. The model is then defined by an initial phase of unstable growth, with subsequent steady state, when the growth becomes constant.

Such a model incorporates the benefits of a careful distinction between the initial growth of the company in its early years, and a growth that applies in the long term. Moreover, the model can be adapted to cases where the firm is expected to post low or even negative growth rates for a few years and then revert back to stable growth.

The initial stage of the model is built on a growth rate that lasts for a set number $T$ of years. After that, the company is assumed to have reached a steady state, and it is possible to model the growth through a perpetuity on the value of future dividends. The value of the stock according to the model is given by

$$\begin{aligned} P_0 &= \sum_{t=1}^{T} \frac{D_t}{1 + r_e^*} + \frac{P_T}{(1 + r_e^*)^T} \\ &= \sum_{t=1}^{T} \frac{D_t}{1 + r_e^*} + \left[ \frac{1}{(1 + r_e^*)^T} \times \frac{D_{T+1}}{(r_e - g)} \right] \end{aligned} \tag{2.2}$$

where

$r_e^*$ is the cost of equity in the high growth stage
$r_e$ is the cost of equity in the stable growth stage

If the initial growth is stable, and the dividend payout ratio is stable too over the $n$ years of extraordinary growth, Eq. (2.2) can be simplified as

$$P_0 = \frac{D_0(1 + g^*)\left[1 - \frac{(1-g^*)^T}{(1-r_e^*)^T}\right]}{(r_e^* - g^*)} + \frac{D_{T+1}}{(r_e - g)(1 + r_e)^T}$$

where

$g^*$ is the growth in the high growth stage

*Example 2.5* Ashare of common stock currently pays a dividend of €1.75 per share and is expected to grow at a rate of 2.25% per year for 2 years and at a rate of 1.75% per year afterwards. The required rate of return is 10.05%. The stock price can be calculated as

$$P_0 = \left[\frac{1.75(1.0225)}{1+0.105} + \frac{1.75(1.0225)^2}{(1+0.105)^2}\right] + \frac{P_2}{(1+0.105)^2}$$

The term $P_2$ is calculated as

$$P_2 = \frac{1.75(1.0225)^2(1.0175)}{0.105 - 0.0175} = €21.28$$

so that

$$P_0 = 1.619 + 1.498 + 17.43 = €20.55$$

The resulting model is very similar to the standard Dividend valuation model, with the difference that in the former case the price obtained is the minimum after which the company experiences constant growth, while in the latter the dividends in the first growth phase are instead discounted using basic cash flow discounting.

It is possible to extend the model to a three-stage formula, by allowing for an initial period of growth, a transitional period where growth declines and a final stable growth phase. It is the most general of the models because it does not impose any restrictions on the payout ratio.

The high growth in stage one is normally higher than classical stable growth, and it is followed by declining rate in the second stage, to then stabilize again at a constant growth rate in the final stage.

The value of the stock is given by discounting the expected dividends during the high growth and the transitional periods, plus the terminal price at the start of the final stable growth phase.

$$P_0 = \sum_{t=1}^{T_1} \frac{EPS_0(1+g_{HG})^t \theta_{HG}}{(1+r_{HG})^t} + \sum_{t=1}^{T_2} \frac{DPS_0}{(1+r_{TR})^t} + \frac{EPS_{T_2}(1+g_{SG})^t \theta_{SG}}{(r_{SG}-g_{SG})(1+r_f)^t}$$

where

$EPS_0$ are the earnings per share in year $t$
$DPS_t$ are the dividends per share in year $t$
$g_{HG}$ is the growth rate in high growth phase, lasting $T_1$ years
$g_{SG}$ is the growth rate in stable phase

$\theta_{HG}$ is the payout ratio in high growth phase
$\theta_{SG}$ is the payout ratio in stable growth phase
$r_{HG}$ is the cost of equity in high growth phase
$r_{TR}$ is the cost of equity in transition phase
$r_{SG}$ is the cost of equity in stable growth phase

As opposed to other DDMs this is much more flexible, with the downturn of requiring more inputs to run. Thus there is a trade-off between flexibility and higher noise in the estimation of the values due to possible errors in the inputs.

Companies experiencing a real three-stage growth process during their life are quite common. Most companies go through a development stage with high growth, a maturing phase with moderate growth, and a declining phase with little, no, or negative growth.

The application of a three-stage dividend discount model is less easy than classical dividend discount models, and it entails six steps, as we can see from the following example.

Consider the valuation of a stock that has a current dividend of €3.00 per share. Dividends are expected to grow at a rate of 12% for the next five years. Following that, the dividends are expected to grow at a rate of 8% for five years.

After ten years, the dividends are expected to grow at a rate of 3% per year, forever. If the required rate of return is 15%, it is possible to calculate the value of the stock by breaking the calculation into six steps.

First of all, the dividends for the years from 1 to 11 must be calculated as

| Year | Dividend growth rate (%) | Dividend |
|---|---|---|
| 1 | 12 | 3.360 |
| 2 | 12 | 3.763 |
| 3 | 12 | 4.215 |
| 4 | 12 | 4.721 |
| 5 | 12 | 5.287 |
| 6 | 8 | 5.710 |
| 7 | 8 | 6.167 |
| 8 | 8 | 6.660 |
| 9 | 8 | 7.193 |
| 10 | 8 | 7.768 |
| 11 | 3 | 8.001 |

| Year | Dividend | Present value |
|---|---|---|
| 1 | 3.36 | 2922 |
| 2 | 3.763 | 2845 |
| 3 | 4.215 | 2771 |
| 4 | 4.721 | 2699 |
| 5 | 5.287 | 2629 |

(continued)

(continued)

| Year | Dividend | Present value |
|---|---|---|
| 6 | 5.71 | 2469 |
| 7 | 6.167 | 2318 |
| 8 | 6.66 | 2177 |
| 9 | 7.193 | 2045 |
| 10 | 7.768 | 1920 |
| 11 | 8.001 | 1720 |

Present value of dividends after year 10 can be calculated as the present value of a growing perpetuity

$$PV_{\infty} = \frac{8.001}{0.15 - 0.03} = €66.67$$

The discounted value at time zero of the growing perpetuity is then given by

$$PV_0 = \frac{66.675}{(1 + 0.15)^{10}} = €16.48$$

The sum of the present value of the dividends in the first 10 years is given by

$$PV_{10} = \sum_{t=1}^{10} \frac{DIV_t}{(1 + 0.15)^t} = €24.79$$

The value of the stock at present is given by the sum of the present value of the growing perpetuity and the present value of the dividends in the first 10 years, as for

$$PV_{\infty} = 16.48 + 24.79 = €41.27$$

It is possible to compare the model to the CAPM in order to grasp the differences among them. The dividend growth model gives a measure of cost of equity through the analysis of empirical data publicly available for most companies.

The calculation comes straight through an algorithm that involves measuring the dividends, estimating the dividend growth, copying the market value of the shares, and using the amounts in the equation to estimate the cost of equity.

The model is however limited in that it does not give any information about why different shares have different costs of equity. This is due to the fact that dividend growth models ignore the risk aspect of valuation.

That model simply measures what's there without offering an explanation. Note particularly that a business cannot alter its cost of equity by changing its dividends. The equation:

$$r_e = \frac{D_0(1+g)}{P_0} + g$$

might suggest that the rate of return would be lowered if the company reduced its dividends or the growth rate.

The reality is different, and normally a dividend cut or growth decrease would result in a lower market value of the company. The value would decrease until the level corresponding to the point where investors obtain the required return.

The CAPM is a more complete model, making a step further by introducing systematic (market) risk in the equation for valuation. Other returns in the economy as well as the relationship among the various risks translate into beta and asset return.

Another important feature of CAPM is to offer several ways to measure the inputs, as the risk-free rate, the market return and the beta. They can be estimated from empirical data, or are normally available as public information.

## 2.4 Dividend Valuation Models and Market Efficiency

The dividend discount model class provides a framework to relate the value of the company to the fundamental features of the business. An example of this relationship is the association between the company's stock's price-to earnings ratio to fundamental factor.

The price-earnings ratio is the ratio of the price per share to the earnings per share of a stock. We can relate this ratio to the company's dividend payout, expected growth, and the required rate of return.

An interesting formula is obtained by taking the DVM equation and dividing both sides by earnings per share. This expresses the price to earnings ratio in terms of dividend payout, required rate of return, and growth, as

$$\frac{P_0}{X_0} = \frac{\frac{D_0}{X_0}(1+g)}{r-g}$$

where

$P_0$ is the current stock price
$X_0$ is the earnings per share
$D_0$ is the current dividend per share
$g$ is the expected dividend growth rate
$r$ is the required rate of return (cost of equity)
$\frac{D_0}{X_0}$ is the dividend payout ratio

The above equation gives important information regarding the relationship between the payout ratio and the price to book equity (PE) ratio. In particular it is clear that the PE ratio is directly related to the dividend payout, inversely related to the cost of equity and directly related to the growth rate of dividends.

It is also possible to rearrange the model to solve for the required rate of return, by noting that

$$P_0 = \frac{D_1}{r-g} \rightarrow r = \frac{D_1}{P_0} + g$$

The formula tells that the required rate of return is composed by the dividend yield and the growth rate of dividends over time, which can be considered as a form of capital gain.

Another way to use the dividend valuation model is to relate the PE ratio to factors such as the dividend payout ratio and the return on equity. Initially one must substitute for the dividend payout ratio in the dividend discount model, so as to obtain

$$P_0 = \frac{D_0(1+g)}{r-g} = \frac{X_0\left(\frac{D_0}{X_0}\right)(1+g)}{r-g}$$

The formula, after manipulation, becomes

$$P_0 = \frac{E_0 \times ROE_0\left(\frac{D_0}{X_0}\right)(1+g)}{r-g}$$

where

$E_0$ is the current book value of equity
$ROE_0$ is the current return on equity

It is then possible to relate the price of a stock to book value, the return on equity, the dividend payout, the required rate of return, and the growth rate. In particular the value of the stock increases when the book value of equity, ROE and growth rate of dividends increases. On the other hand the value decreases when the rate of return applied to discount the dividends increases.

The dividend discount model can in conclusion be used, along with our knowledge of financial relations (i.e., financial statements and financial ratios), to relate the stock's price and price multiples to fundamental factors.

Stock valuation models are based on the belief that the investors will act rationally by paying as a price for a stock the present value of the benefits expected to be gained in the future.

But in reality stock valuation is not so simple, because reality shows that dividends are not paid constantly, nor do they grow at a constant rate for most companies. The model doesn't apply in the case when dividends do not grow at a constant rate (or at least in stages) or in the case when the company does not pay dividends.

Those cases must be handled with the use of other models, based on other types of cash flows, as shown in the following chapters of this book. When valuing investments, the analyst wants to estimate the future cash flows from these investments and then discount these to the present.

The logic behind this process is that no one will pay more than logically expected as a fair price on any investment. In an efficient market this means that the price today reflects all available information.

The information itself relates to the amount of future cash flows and the risk associated to them. As the market is hit by new pieces of information, the stock price will change according to the reaction of the investors.

In terms of valuation, the more complex the information and valuation of the information, the more time it takes for the market to digest the information and the stock to be properly valued.

For well established companies the new pieces of information will display their effect in about fifteen minutes, so that the normal individual investor will never be quick enough to benefit from the changes.

Technical analysis will therefore be useless, and fundamental analysis, while valuable in terms of evaluating future cash flows, assessing risk, and assisting in the proper selection of investments for a portfolio, will simply produce returns commensurate with the risk assumed.

## Bibliography

Aharony J, Swary I (1980) Quarterly dividend and earnings announcements and stockholders' returns: an empirical analysis. J Finance 35:1–12

Black F, Scholes M (1974) The effects of dividend yield and dividend policy on common stock prices and returns. J Financ Econ 1:1–22

Gordon MJ (1959) Dividends, earnings and stock prices. Rev Econ Stat 41(2):99–105 (The MIT Press)

Lintner J (1964) Optimal dividends and corporate growth under uncertainty. Q J Econ 78:49–95

# Chapter 3
# Free-Cash-Flow-Based Methods

**Abstract** According to what is the target of the valuation, both free cash flows to the firm and free cash flows to equity can be calculated. When the cash flows have been calculated the valuation comes from the application of the Free-Cash-Flow-Based valuation methods, a family of models that relies on the accounting cash flows as a source of information for the value calculation. A further step entails the estimation of a growth rate for the cash flows in the future, so to adjust the model for a more realistic framework. After the growth rate is estimated, the valuation principles follow those of dividend discount models. It is possible in fact to set up single-stage or multi-stage models for both asset and equity valuation. A discrete sequence of discrete cash flows to be discounted at present value, together with a terminal value, represents the most popular approach to corporate valuation.

**Keywords** Income · Depreciation · Debt · Working capital · Capital expenditures · Free cash-flow · Equity · WACC

The free cash flow methods are an important section of the company valuation theory and methods. It relies on the cash flows generated by the corporate activity as a base for value calculation.

As oppose as dividends, the free cash flows to the firm, or to equity, are directly obtained by reworking the entries in the income statement, in order to isolate the cash flows that are generated by a specific investment or the company as a whole.

Different types of cash flows require different types of discount rate to be moved back to present value, in order to be summed up and give the value of the company's assets or equity.

It is therefore important to learn how the cash flows can be calculated, and how the other inputs analysed in the rest of the book mix together to obtain a reliable estimation of the corporate value.



A. Corelli, *Inside Company Valuation*, SpringerBriefs in Finance,
DOI 10.1007/978-3-319-53783-2_3

After studying this chapter, you will be able to answer the following questions, among others:

1. How can the free cash flows to the firm be calculated?
2. How can the free cash flows to the equity be calculated?
3. How can the growth rate of cash flows over time be estimated?
4. What is the relationship between different types of cash flows and the value of the firm?
5. What are the main steps for valuation based on cash flows?

The first section of the chapter is about the calculation of cash flows of several types, and the difference between them. The second section is about the estimation of the growth rate of cash flows over time. The third section focuses on the application of the different types of cash flows to company valuation.

## 3.1 The Determinants of Cash Flows

The models of free cash flows try to analyze the amount of cash that shareholders can get as a return on their investment, as a result of the company operations. They are calculated starting from the net income.

Net income measures the earnings of the company over some period, and in order to split from it the part of income that does not belong to shareholders, it is necessary to convert to a cash flow by subtracting out a firm's reinvestment needs.

The road to cash flows goes through netting the net income from various expenses further to those related to production. For example, one must subtract any capital expenditures, defined broadly to include acquisitions, since they represent cash outflows.

It is very important to recall that depreciation and amortization, subtracted first from the gross income to get the EBIT, are not cash flows. They must therefore be added back to the net income. Finally, net expenditure is then the difference between capital expenditures and depreciation.

The capital expenditures are normally related to the growth features of the company. For example firms that experience high growth normally have high net capital expenditures, while low-growth firms have low or sometimes even negative net capital expenditures.

An increase in the working capital generates a reduction in cash flows, and vice versa, so that companies with lowest working capital have the highest cash flows available to shareholders.

Companies experiencing a fast growth are subject to large increases in working capital, especially in industries characterized by high levels of working capital, like retailing. In order to capture cash flow effects of working capital one must focus on the non-cash working capital.

The final important factor that affects the cash flows is the leverage of the firm. Shareholders suffer changes in debt, because they impact on their cash flows, and the interest and principal repayment affect the share of cash going to equity.

The payment of debt itself is a cash outflow but can be normally financed by the issuance of new debt, which itself is a cash inflow. The last point in determining cash flow is debt. Equity holders are typically subject to changes in the level of debt, that impact on their cash flows.

It is possible to calculate the cash flows through a formula that includes all variables of interest, like the changes in working capital, net changes in debt, and the net capital expenditures. The result is the free cash flow to equity (FCFE) calculated as

$$\begin{aligned} FCFE =& \, Net\,Income - (Capital\,Expenditures - Depreciation) \\ & - (\Delta\,in\,non - cash\,working\,capital) + (New\,debt\,issued - Debt\,repayments) \end{aligned}$$

The cash flow as calculated above normally goes to shareholders in the form of dividends or it is retained as additional equity to be reinvested. It is possible to assume that net capital expenditures and working capital are financed by a mix of equity and debt capital. The proportion of net capital expenditures and working capital raised in the form of debt can be indicated as $d$, and the effect on cash flows to equity of these items can be represented as

$$\begin{aligned} FCFE =& \, Net\,Income - (Capital\,Expenditures - Depreciation) \\ & - (\Delta\,in\,non - cash\,working\,capital) + (New\,debt\,issued - Debt\,repayments) \end{aligned}$$

It is possible to define the cash flows associated with capital expenditure needs $(FCFE_{CAP})$ as

$$FCFE_{CAP} = -(Capital\,Expenditures - Depreciation)(1 - d)$$

On the other hand, we can also define the cash flows associated with working capital needs $(FCFE_{WC})$ as

$$FCFE_{WC} = -(\Delta\,in\,non - cash\,working\,capital)(1 - d)$$

It is then possible to combine the above definitions and formulas to express the cash flow to equity for a constant debt-to-equity ratio, as

$$\begin{aligned} FCFE =& \, Net\,Income - (Capital\,Expenditures - Depreciation)(1 - d) \\ & - (\Delta\,in\,non - cash\,working\,capital)(1 - d) \end{aligned} \tag{3.1}$$

As it shows by Eq. (3.1), the debt is not included in the formula anymore, because the new debt issued is financing the previous debt, and the debt-equity ratio stays constant, as the model assumes.

As from Corelli (2016) it is particularly useful to assume that a specified proportion of net capital expenditures and working capital needs will be financed with debt if the target or optimal debt ratio of the firm is used to forecast the free cash flow to equity that will be available in future periods. For past periods, an alternative approach is to use the firm's average debt-to-equity ratio over the period to arrive at approximate free cash flows to equity.

In the traditional model of cash flow valuation the underlying assumption is that no preferred dividends are paid. The model then values the common equity and if one wants to implement the model for preferred dividends, the equation must be modified.

The preferred dividends must be subtracted from the net income as well, in order to arrive at the FCFE, according to the modified formula

$$\begin{aligned} FCFE =& Net\, Income - (Capital\, Expenditures - Depreciation) \\ & - (\Delta\, in\, non - cash\, working\, capital) - (Pref.\, dividend + New\, pref.\, stock\, issued) \\ & + (New\, debt\, issued - Debt\, repayments) \end{aligned}$$

which, in the short form, becomes

$$\begin{aligned} FCFE =& Net\, Income - Pref.\, dividends - (Capital\, Expenditures - Depreciation)(1 - d) \\ & - (\Delta\, in\, non - cash\, working\, capital)(1 - d) \end{aligned}$$

In this case the non-equity financial ratio ($d$) also includes the expected financing from new preferred stock issues.

In the end there is not much difference between standard FCFE model and the dividend discount model. One can see the FCFE model as a framework where potential rather than actual dividends are discounted.

The various versions of the model based on cash flow are therefore all variants of the dividend discount model, with a major change given by the fact that the dividends are replaced by the free cash flows to equity.

Such a replacement goes beyond changing one type of cash flow with another, given that it entails the assumption that the FCFE will be paid out to stockholders. Two logical consequences follow.

As a consequence of the above, if all cash is paid out there will be no retained equity in the company, and only the little reinvested part will not go to the shareholders, therefore draining the liquidity from income.

Also, the part of growth in income that comes from the current assets will not be part of the expected growth in FCFE, and only the net income from operating asset will be contributing to that growth.

The model gets more complicated when the stock repurchases are included in the dividend amount to be discounted. It is therefore useful to analyze how the FCFE model compares to the modified dividend model in this case.

The stock repurchases are accumulation of cash due to the managerial choice to not pay dividends, but retain the cash instead. Given that, the FCFE can be seen as a proxy of the return che companies can offer to the shareholders over time as dividends or stocks repurchase.

## 3.2 The Growth Rate

The growth of cash flows can be estimated by using the same approach used for the dividends, given they both represent cash flows to equity investors. The expected growth rate can then be expressed as

$$E(g) = R_R \times r_e$$

where

$R_R$ is the retention ratio

The amount of cash flow that is not paid out as dividends is reinvested into the firm as retained earnings. This is not consistent with the assumption that free cash flows to equity are paid out to stockholders which underlies FCFE models.

For a better fit with reality, the model needs to be modified by replacing the retention ratio with the equity reinvestment rate, which measures the share of net income that is reinvested into the corporate activities. The model can be then formulated as

$$R_E = 1 - \frac{Net\ capital\ expenditure + \Delta\ in\ working\ capital - Net\ debt\ issues}{Net\ income}$$

One must acknowledge that the FCFE model does not account for any extra cash available to the firm after payments are made to the shareholders. It is therefore appropriate to use an equity discount rate that does not include the interest income from current assets.

A similar argument works for the book value of equity, in the part which is made of the value of cash and marketable securities. It is then necessary to build a new version of the return on equity that measures the non-cash aspects, as

$$Non - cash\ ROE = 1 - \frac{Net\ income - After\ tax\ income\ from\ cash\ and\ marketable\ securities}{Book\ value\ of\ equity - Cash\ and\ marketable\ securities}$$

The product of the equity reinvestment rate and the modified ROE will yield the expected growth rate in FCFE as

$$Expected\ Growth\ in\ FCFE = Equity\ reinvestment\ rate \times Non - cash\ ROE$$

Depending on the type of company to valuate, there are many versions of the FCFE that can be used. For example, for a company that is growing at a stable rate, the best model to use is the constant growth FCFE model.

Future values of the FCFE are discounted in order to get the value of equity, given the stable growth rate and the required rate of return.

$$P_0 = \frac{FCFE_1}{r_e - g}$$

where

$FCFE_1$ is the free cash flows to equity in period 1.

In case of a constant growth company, the model is based on assumptions that are very similar to those of the Gordon dividend growth model. The analysis must be based on a reasonable growth rate.

In particular the permanent growth rate should never exceed the growth rate in the economy where the company operates. Other assumptions included in the model are peculiar of stable companies in general.

An example of a firm-specific assumption is that stable companies have an average risk, given the amount of capital expenditures being not too large relative to depreciation. A consequence of that is that when CAPM is used for determining the cost of equity, the beta should be not significantly different from one.

The reinvestment rate for a stable growth firm can be estimated in two ways. A first approach is to use the typical reinvestment rates for firms in the industry of operation. Alternatively, the relationship between growth and fundamentals can be used to estimate the required reinvestment.

The expected growth in net income can be written as:

$$Expected\ growth\ rate\ in\ net\ income = Equity\ Reinvestment\ Rate \times Return\ on\ equity$$

which turns in an estimation of the equity reinvestment rate as

$$Equity\ Reinvestment\ Rate = \frac{Expected\ growth\ rate\ in\ net\ income}{Return\ on\ equity}$$

As an example, consider a firm with stable growth rate of 3% and a cost of equity of 9%. With such numbers, the company needs to reinvest the 33.3% of its net income in order to finance capital expenditures and working capital needs. In this way, the FCFE should be the 66.7% of the net income.

The FCF valuation, either done with cash flows to equity or cash flows to firm is suitable in some specific cases which can be listed as:

- The company is not dividend-paying.
- The company is dividend paying but dividends significantly differ from fcfe.

- The company's FCF's align with company's profitability within a reasonable time horizon.
- The investor has a control perspective.

Recall that the maturity of a stock is infinite as the life of the company is assumed to be. Valuation must take this into consideration and it is therefore appropriate to divide the process in two or more stages.

Valuation based on a 2-stage procedure underlines that there are $N$ future years (usually 5–10) of growth forecast of cash flows, followed by the final (discounted) terminal value.

$$V_0 = \sum_{i=1}^{N} \frac{FCF_i}{(1 + r_{WACC})^i} + \frac{TV}{(1 + r_{WACC})^N}$$

As mentioned before, the permanent growth rate used to calculate the terminal value should be never higher than the growth rate of the economy in which the company is operating. A good proxy for it is the long-term risk-free rate in the economy.

When nominal cash flows and discount rates are used, the growth rate should be nominal in the currency in which the valuation is denominated. One simple proxy for the nominal growth rate of the economy is the risk-free rate.

In order to estimate the growth, it is normally appropriate to start from the historical growth in earnings. Normally analysts do EPS analysis for a range of firms, in an industry, which is useful to know what the estimates are. It is possible to trace all growth in earnings to the amounts invested in new projects by the company, and what returns these projects are making for the firm.

Growth estimation can be performed by analysing the EPS of the firm. There are two possible methods for it. The first method assumes that the ROE is expected to stay constant and defines the growth rate as

$$g_{EPS} = r_R \times ROE$$

Please note that in no way a company can sustain a growth in earnings higher than its ROE in the long term. When the ROE is expected to change, the equation becomes

$$g_{EPS} = (r_R \times ROE_{t+1}) + \frac{(ROE_{t+1} - ROE_t)}{ROE_t}$$

Please recall that small changes in ROE translate into large changes in the expected growth rate, and that the lower the current ROE, the greater the effect on growth of changes in the ROE.

There are a few simple rules to be taken into account when estimating the stable growth. The first rule is that a long-term growth cannot be ever sustained by any firm by just improving the ROE.

At the stable growth state the company should show the features of comparable steady companies. In particular, the risk of the firm, as measured by beta and ratings, should reflect that of a stable growth firm.

The reinvestment rate of the firm should reflect the expected growth rate and the firm's return on capital

$$r_R = \frac{g_{OC}}{ROC}$$

In case the cash flows can be regarded as growing at a constant rate forever, the calculation of the value becomes a growing perpetuity as already seen in previous chapters.

Normally analysts assume growth periods to be long and characterized by substantial excess returns, the evidence suggests that they are much too optimistic. A study of revenue growth at firms that make IPOs in the years after the IPO shows that typically the revenue growth rate of a new public company outpaces its industry average for only about five years.

The general method for cash flow discounting can be summarized in formula as

$$V = \frac{C_1}{1+r} + \frac{C_2}{(1+r)^2} + \ldots + \frac{C_n + RV_n}{(1+r)^n}$$

where

$C_i$ is the cash flow generated by the company at time $i$
$RV_n$ is the residual value of the company at time $n$
$r$ is the appropriate cost of capital (risk-weighted discount rate)

The series of cash flows in the above formula appears to be discrete, but in reality the residual value $RV_n$ is calculated as a perpetuity calculated on the cash flows following the first period.

A simplified procedure for considering an indefinite duration of future flows after the year $n$ is to assume a constant growth rate $g$ of flows after that period. Then the residual value in year $n$ is

$$RV_n = \frac{RV_n \times (1+g)}{r-g}$$

The perpetuity formula per se ignores the effect of the cash flows ahead of time, as a mathematical tool. This is also acceptable from the point of view of the firm's valuation in that their present value decreases progressively with longer time horizons.

The first thing to do before entering into the detailed steps of cash flow based corporate valuation is to define the several types of cash flows that can be used in the analysis, and how they can be used.

## 3.3 Cash Flows and Corporate Value

Three types of cash flows can be calculated. The debt cash flow is the easiest to understand and is the sum of the interest payments to be repaid on debt, together with the principal payments.

In order to determine the present market value of the existing debt, this flow must be discounted at the required rate of return to debt (cost of the debt). The market value of the corporate debt in most cases will be equal to its book value, which is often considered as a fair approximation of the former.

The free cash flow to the firm (FCF) is used to obtain the company's total value, including equity and debt. The FCFE enables the value of the equity to be obtained, which, combined with the value of the debt, will also enable the company's total value to be determined.

Recall that the company's financial assets consist of the net fixed assets plus the working capital requirements. The financial liabilities on the other hand consist of the equity in the hands of the shareholders, plus the total debt.

Good company valuation requires forecasting cash flows further ahead in time than is normally done in any cash budget. This information cannot be supplied by the accounting given the use of accrual approach and allocation of revenues, costs and expenses using basically arbitrary mechanisms.

The appropriate approach for cash flow calculation is the cash approach, relating to the actual flows of cash coming in or going out the company. However, when the accounting is adjusted to this approach, we can calculate whatever cash flow we are interested in.

Consider an hypothetical company ABC, of which one wants to analyse the financial statements. The information given in the accounting statements must be adjusted to give the cash flows for each period.

The company's free cash flow by definition must not include any payments to fund providers. Therefore, dividends and interest expenses must not be included in the free cash flow.

It is necessary to directly calculate the EBIT and the tax payable on it, so as to obtain the net income before subtracting any interest payment. Then one must add the depreciation for the period because it is not a payment but merely an accounting entry.

Finally, the analysis must account for the sums of money to be allocated to new investments in fixed assets and new working capital requirements (WCR), as these sums must be deducted in order to calculate the free cash flow (Tables 3.1 and 3.2).

The point with cash flow valuation is to ignore the financing sources for the company and focus on the financial return on the company, net of taxes. The view is dynamic and takes into account the perspective of the ongoing activities. If the company is not leveraged, with zero debt. the free cash flow are equal to the equity cash flow, which is another cash flow variant used in valuations.

**Table 3.1** An example of corporate net income

| €M | 2013 | 2014 | 2015 |
|---|---|---|---|
| Sales | 100 | 110 | 121 |
| Cost of sales | (65) | (71.50) | (78.65) |
| General expenses | (18.90) | (20.79) | (22.87) |
| Depreciation | (2) | (2) | (2) |
| *EBIT* | *14.10* | *15.71* | *17.48* |
| Interest expenses | (1) | (1) | (1) |
| *EBT* | *13.10* | *14.71* | *16.48* |
| Taxes | (4.585) | (5.149) | (5.768) |
| *Net income* | *8.515* | *9.562* | *10.71* |
| Dividends | (3.406) | (3.825) | (4.285) |
| *Retained earnings* | *5.109* | *5.737* | *6.428* |

**Table 3.2** An example of corporate cash flow statement

| €M | 2013 | 2014 | 2015 |
|---|---|---|---|
| *EBIT* | *14.1* | *15.71* | *17.48* |
| Taxes on EBIT | (4.94) | (5.50) | (6.12) |
| Net income without debt | 9.165 | 10.21 | 11.36 |
| Depreciation | 2 | 2 | 2 |
| Change in fixed assets | (6.10) | (6.71) | (7.38) |
| Change in working capital | (1.10) | (1.21) | (1.33) |
| *EBT* | *3.965* | *4.292* | *4.651* |

The valuation done with FCFE gives the value of the company's equity. It is therefore appropriate to discount the cash flows using the cost of (return on) equity. The company total value is then the sum of the calculated equity and the value of debt.

The valuation can be also done in terms of the sum of the unlevered value of the firm and the present value of the corporate tax shield, which is the monetary tax advantages coming from the fact that interest payments on debt are tax deductible.

The value of the company without debt is obtained by discounting the free cash flow, using the rate of required return to equity that would be applicable to the company if it were to be considered as having no debt.

The appropriate rate to discount the cash flows of the unlevered firm is the unlevered cost of capital, which correspond to the required rate of return to equity if the company had debt in its capital structure.

The figure underlines the fact that a shareholder of the company would bear the financial risk implied by the existence of debt and would demand a higher equity risk premium.

In case the company is not leveraged, the required return of equity corresponds to the weighted average cost of capital (WACC), and equity is the only source of capital in the firm's capital structure.

When the company is leveraged, the tax shield arise from the lower taxes paid due to the interest paid on the debt in each period. In order to find the present value

**Table 3.3** A scheme of company valuation steps

| | |
|---|---|
| **1. Analysis of the company** | |
| Financial analysis | Strategic analysis |
| Financial statements | Industry |
| Business risk | Competitiveness |
| **2. Projection of future cash flows** | |
| Financial forecasts | Strategic forecast |
| Projected cash flows | Industry projections |
| Terminal value | Consistency |
| **3. Determination of the cost of capital** | |
| Cost of capital | WACC |
| Cost of equity | CAPM |
| Cost of debt | Growth |
| **4. Net present value of future cash flows** | |
| **5. Interpretation of the results** | |
| Comparable companies | Sensitivity analysis |
| Benchmarking | Time horizon |

of the tax shield, we would first have to calculate the saving obtained by this means for each of the years, multiplying the interest payable on the debt by the tax rate.

The obtained cash flows must be discounted at the appropriate discount rate. There are diverse opinions about what discount rate to use in this case. Some analysts suggest using the debt's market cost, which need not necessarily be the interest rate at which the company has contracted its debt.

Valuation through cash flows can be finally summarized through the steps reported in Table 3.3.

## Bibliography

Beranek W, Howe KM (1990) The regulated firm and the DCF model: some lessons from financial theory. J Regul Econ 193

Corelli A (2016) Analytical Corporate Finance, Springer

Rosenberg B, Guy J (1976) Prediction of beta from investment fundamentals: part one, prediction criteria. Financ Anal J 32(3):60–72

Ross SA, Westerfield R, Jordan BD (2008) Corporate finance fundamentals. McGraw-Hill Irwin, New York

Stulz RM (2000) Globalization, corporate finance, and the cost of capital. J Appl Corp Finance 12

# Chapter 4
# The Valuation of Private Firms

**Abstract** The absence of time series and other public data about a private company forces the analyst to estimate some of the parameters required to carry on the valuation. In particular the beta, cost of debt and growth rate are needed for the purpose. When the information about comparable firms is acquired, it is then possible to estimate the beta as an average, and use such a figure as a good proxy of the beta of the private firm. The same holds for the cost of debt, especially in case the publicly traded comparable firms issue bonds and have a publicly observable cost of debt to be used again as a proxy of the figure for the private firm. The growth rate of the cash flows overtime is a factor that becomes crucial in the case of private firms. The financial history of the corporation helps in achieving good estimates.

**Keywords** Beta · CAPM · Leverage · Cost of debt · Terminal value · Long-term growth

Private companies differ from publicly traded ones in that their shares are not traded on public markets, but they belong to private investors. Knowledge on how to perform valuation of those companies is a primary achievement in corporate finance.

Following the previous chapters, the aim is to identify and learn the tools and techniques to value a private firm based on inputs that are necessarily not the same as the ones used to value public corporations.

In this sense it is very important to understand the dynamics of cash flows, growth and discounting rates in the framework of absence of public data. The only source of information in this case are the figures included in the financial statements.

The analyst is challenged with achieving a good valuation in tough conditions, without having access to time series of any sort. This makes it more complicated to estimate parameters like beta, cost of debt, and growth rate, that are essential to the calculation of the cost of equity and cost of capital.



A. Corelli, *Inside Company Valuation*, SpringerBriefs in Finance,
DOI 10.1007/978-3-319-53783-2_4

It comes without saying that performing the valuation of a private company is of primary interest when an IPO takes place. Even if the valuation of a private company can be performed at any stage of its life, it is undoubtful that the IPO process is the one that mostly benefits from such a task.

After studying this chapter, you will be able to answer the following questions, among others:

1. How can the CAPM be calculated for private firms?
2. How can its parameters of CAPM be estimated in absence of public data?
3. What is the process for estimating the cost of debt for a private company?
4. How can a reasonable growth rate be estimated for the cash flows of a private company?
5. How is the calculation of the terminal value and the present value of cash flows influenced by the nature of the private firm?

The first section of the chapter is about the estimation of the CAPM beta coefficient, so as to apply it in practice for cost of equity estimation. Also the market return and beta estimation are described. The second section is about the estimation of the cost of debt. The third section focuses on the estimation of growth rates and the calculation of the terminal value.

## 4.1 Beta Estimation

When it comes to private companies, the valuation becomes a different task compared to the valuation of a public firm. The major issue is about what model to choose and whether to focus on the value of the company or the value of the equity.

On top of that the growth stage of the company is an important factor, and according to the seniority of the company the choice will be between a steady state model and a two-stage or three-stage model.

The difficulties in the estimation of the right discount rate to be used to discount the cash flows come as a consequence of the above choice. The choice of either cost of equity, cost of debt, or overall cost of capital must be consistent on what type of cash flows have to be discounted.

There are several challenges in estimating the cost of equity for a private firm. Classical models like CAPM can be applied but one should recall that they are based on market data.

Market data are obviously not available for a private company and it is therefore necessary to develop methods for the estimation of the model parameters, in order to overcome the limitations imposed by the lack of data.

One task to be accomplished is the estimation of the beta to be used in the CAPM calculation of the cost of equity. In order to solve the task properly, in absence of market data to estimate it, one has to use comparable firms.

The firms to be used as a bunch of comparable traded companies, to be used for beta estimation should preferably be in the same line of business, but more generally, affected by the same economic forces that affect the firm being valued.

An initial test of compatibility of the selected companies with the target company can be done by estimating the correlation between the financials of the firms. In the event of an estimated high correlation, the comparable firms can be used for the analysis.

The second step is to estimate the average beta for the publicly traded comparable firms. After that, the average market value of the debt-to-equity ratio for the comparable firms must be estimated. This leads to the calculation of the unlevered beta of the firm.

$$\beta_U = \frac{\beta_L}{1 + (1 - \tau_C)\left(\frac{D}{E}\right)}$$

where

$\beta_U$ is the unlevered beta
$\beta_L$ is the levered beta
$\tau_C$ is the corporate tax rate
$D$ is the market value of the debt of the firm
$E$ is the market value of the equity of the firm

How to estimate the debt-to-equity ratio for a private firm is another challenging task, given that only book values are available for both. It is again possible to use figures based on comparable firms in the industry.

Following Corelli (2016), assuming that the private firm will move to the industry average debt ratio, the beta for the private firm will then also converge on the industry average beta. This might not happen immediately but over the long term.

$$\tilde{\beta}_L = \frac{\beta_U}{1 + (1 - \tau_C)\left(\frac{D_{AVG}}{E}\right)}$$

where

$\beta_{PVT}$ is the beta of the private firm
$D_{AVG}$ is the industry average debt

Another approach involves estimating the theoretical optimal debt-to-equity ratio for the private firm, based on the analysis of the financial statement. The determined ratio can then be used for the beta calculation. One should consider to then use the same ratio also in the estimation of cash flows.

$$\tilde{\beta}_L = \frac{\beta_U}{1 + (1 - \tau_C)\left(\frac{D_{OPT}}{E}\right)}$$

where

$D_{OPT}$ is the optimal debt

The final step of this method is to use the beta obtained through the above steps to estimate the cost of equity of the private firm.

An alternative method consists in estimating an accounting beta, again with a process divided in several steps. First of all, the historical accounting earnings of the private firm must be collected for a reasonably long time.

Collecting the accounting earnings for the index of reference, as a proxy of the market, for the same time period is the following step. The changes in earnings for the private company must then be regressed against the changes in the market returns. The beta is then obtained as the slope of the regression line.

Much research has been done by researchers to relate the betas of publicly traded firms to observable business variables. Fundamental betas are the result of the efforts of authors like Beaver et al. (1970).

The above authors have related the betas with several corporate variables including leverage, liquidity, asset size and earnings. The betas have been regressed on the variables in order to find the appropriate coefficients to relate the beta for the private firm to its own business variables.

It is possible to estimate the beta for the firm by looking at the coefficients of each variable, in a framework that is easy to implement, but limited in efficiency to the significance of the underlying regression.

It is also possible to estimate another type of beta, the bottom-up, by looking at competiting companies in the same business sector. The betas of the companies are then averaged based on their weights in the industry and the resulting number is adjusted for leverage.

The four main steps involved in the estimation of bottom-up beta are:

- Break the company down into the businesses that it operates in. The business should not be defined too narrowly in order for the process to run smoothly.
- Estimate the risk (beta) of being in each business. This beta is called an asset beta or an unlevered beta.
- Take a weighted average of the unlevered betas of the businesses, weighted by how much value the company gets from each business.
- Adjust the beta for the company's financial leverage (Debt to equity ratio).

In order to adjust the bottom-up beta to the leverage of the firm, there is a controversy related to the fact that normally such an adjustment should be done by using market values of equity and debt.

Normally, for a publicly traded company, by regressing the asset premium against the market premium, beta can be obtained as the coefficient on the regressor. The following equation is employed to estimate the beta of the company:

$$\Delta S_i = \alpha_i + \beta_i \Delta S_M + e_i$$

where

$\Delta S_i$ is the change in price of stock $i$
$\Delta S_M$ is the change in the value of the market
$\alpha_i$ is the intercept value of the regression
$\beta_i$ is the beta of the stock $i$
$e_i$ is the error term

Again the problem of this approach is that it can be only applied to public companies, for which it is possible to get the data needed for the regression. Due to lack of market data on stock prices of private companies, it is not possible to estimate stock beta.

These values are not available for private companies, but there are several alternatives that can be used to overcome the issue. One option is to assume that the market leverage of the private firm resembles the average leverage of the industry sector. In this case the levered beta can be defined as an industry beta.

The estimated beta from comparable firms is a proxy for the industry average levered beta. After that, the beta needs to be unlevered, by applying the average debt-to-equity ratio for these comparable companies. The final step is re-levering beta, using private company's target debt/equity ratio.

*Example 4.1* Consider estimating the beta of an illustrative company with a target debt to equity ratio of 0.5, by using the information about four comparable companies:

| Comparable companies | Beta | Debt | Equity | D/E |
|---|---|---|---|---|
| Company 1 | 1.60 | 7840.00 | 16,267.00 | 0.48 |
| Company 2 | 1.65 | 10,565.00 | 37,850.00 | 0.28 |
| Company 3 | 1.71 | 523.23 | 1653.47 | 0.32 |
| Company 4 | 1.69 | 1627.84 | 4079.74 | 0.40 |
| *Averages* | | | | |
| Weighted average beta | 1.64 | | | |
| Weighted average D/E | 0.34 | | | |

The equity-weighted average beta of the four companies is 1.64. This is close to the arithmetic average of about 1.66. The next step is to unlever the beta by using the average debt-to-equity ratio for these companies, which is 0.34. It follows that

$$\beta_U = \frac{\beta_L}{1 + (1-\tau)\frac{D_{AVG}}{E_{AVG}}} = \frac{1.64}{1 + (1-0.35)0.34} = 1.343$$

where

$\beta_L$ is the levered beta
$\frac{D_{AVG}}{E_{AVG}}$ is the average debt-to-equity ratio of comparable firms
$\tau$ is the tax rate

Thus, we get the unlevered beta of 1.343 approximately.

In the final step, we need to re-lever the equity using the target D/E ratio of the private company, which equals 0.5.

$$\beta_L = \beta_U \left[1 + (1 - \tau)\frac{D}{E}\right] = 1.343(1 + 1 - 0.35)0.5 = 1.78$$

The method of the average beta of comparable firms has the drawback consisting in the fact that it neglects the differences in size between companies. It happens often that the public comparable companies considered are much larger than the private firm.

Another possible method is the Earnings Beta Approach that becomes useful in cases when it is difficult to obtain reliable comparable betas. It happens quite often in fact that public companies operate in many segments, making it difficult to identify which companies are really comparable to the private firm under valuation.

The Earnings Beta Approach uses the company's historical earnings change and regresses them against the market returns. An appropriate market index can be used as a proxy for market.

The beta obtained as above needs to be adjusted in order to reflect the expected future performance of the firm. Recall that beta has a mean-reverting feature, and tends to get to one in the long term. The adjusted beta can be expressed as

$$\beta_{Adj} = \alpha(1 - \alpha)\beta_H$$

where

$\beta_{Adj}$ is the adjusted beta
$\alpha$ is a smoothing factor
$\beta_H$ is the historical beta

The smoothing factor can be derived through complex statistical analysis based on historical data, but as a rule of thumb, the value of 0.33 or (1/3) is used as a proxy.

There are drawbacks in the earnings beta approach, depending on the fact that private companies do not exhibit long time series of earnings, therefore making the regression analysis less reliable.

Moreover, the accounting earnings are subject to smoothing and accounting policy changes. Therefore, these may not be appropriate for statistical analysis, unless necessary adjustments have been made.

The use of CAPM for valuation of private companies can be problematic, due to the lack of a direct method to estimate the beta of a private firm. The above-presented methods represent a good compromise for implementing the model.

It is also possible to use a method that takes into consideration the company's target debt-to-equity ratio and estimate the beta based on it. However it is not easy to apply this method because the ratio is normally disclosed only upon decision of the management in that sense.

It is important to note that the beta represents the risk added by adding some asset or portfolio to an already well diversified portfolio. The owner is often the only investor and thus can be viewed as the marginal investor.

This means that the entrepreneur is not well diversified but concentrates most of his wealth in the private firm. It can be stated as a consequence that the betas calculated for a private firm will generally underestimate the risk exposure.

In case an adjustment for non-diversifiable risk is needed, it is possible to use a method based on the standard deviation by defining the standard deviation in the private firm's equity value by $\sigma_i$ and the standard deviation in the market index by $\sigma_m$.

Define the coefficient of correlation between the single stock and the broad index as $\rho_{im}$. The market beta can then be defined as:

$$\beta_m = \rho_{im} \frac{\sigma_i}{\sigma_m}$$

where

$\sigma_i$ is the volatility of the returns of stock $i$
$\sigma_m$ is the volatility of the market returns
$\rho_{im}$ is the correlation coefficient between market returns and returns of stock $i$

To calculate the total firm's risk, it is possible to divide the market beta by the correlation coefficient $\rho_{im}$ leading to

$$\frac{\beta_m}{\rho_{im}} = \frac{\sigma_i}{\sigma_m} \tag{4.1}$$

Equation (4.1) shows the relative standard deviation of the company, scaled against the market risk. The measure represents the total beta of company's equity, which accounts for non-diversifiable risk.

$$\beta_{TOT} = \frac{\beta_m}{\rho_{im}}$$

As from the equation, the beta depends on the correlation between the firm and the market. The beta will be higher when the correlation coefficient is lower. The calculation of the total beta for a private firm can be difficult given the lack of market prices, which makes it difficult to calculate the correlation coefficient.

However, one should recall that it is possible to estimate the market beta of the sector by looking at publicly traded firms in the business. The correlation coefficient is obtained by looking at the same sample and using it to estimate a total beta for a private firm.

Another important factor affecting the choice for beta estimation is the purpose of the valuation. In case the valuation is done before the sale of the company, the need for adjustment of the beta strictly depends on the potential buyer.

In case of an IPO for example, there is no need for an adjustment in that the market investors representing the buying body, are assumed to be already fully diversified in their portfolio.

If the valuation is for sale to another individual or private business instead, the extent of the adjustment is dependent on the diversification of the buyer. The more diversified the buyer, the higher the correlation with the market and the smaller the total beta adjustment.

Beta is a measure of systematic risk of a security, which compares the excess return (risk premium) on the security itself and the excess return on the market. It also delivers information on how the cost of equity of a company changes, according to changes of the return of the market.

## 4.2 Cost of Debt Estimation

A basic problem in the valuation of private firms is the determination of the cost of debt. As for the calculation of the cost of equity, private firms do not issue publicly traded bonds, and they are therefore not rated.

For most companies the bank debts are the biggest part of the corporate long term debt, as from the balance sheet. However, using the interest rate paid on the loans as a proxy for the cost of debt can be misleading.

A method to overcome the issue is to assume again that an analysis of the comparable firms can give a good estimation of the figures of the target company. In particular, by looking at the market cost of debt of public comparable companies, one can estimate the cost of debt of the target company, as an average. In this case the cost of debt can be expressed as

$$\tilde{r}_D \cong r_{AVG}$$

where

$r_{AVG}$ is the cost of debt for similar firms in the industry

A second method is based on the estimation of the appropriate credit rating of the company, which can be calculated from the financial ratios. Then one can look at the market estimated yields for bonds of the same rating class, and use an average of these as the cost of debt. This leads to a formulation of the type

$$\tilde{r}_D \cong r_{RAT}$$

where

$r_{RAT}$ is the interest rate based upon estimated bond rating for private firm

Finally a third method relates to the current long-term of the private company, and selecting just the long-term interest paid on it as a proxy for the cost of debt. The use of such a method makes sense if the company acquires that debt fairly recently, compared to the time valuation is performed. In this case the formula is

$$\tilde{r}_D \cong \frac{I}{D}$$

where

$I$ is the interest expense

When both cost of equity and cost of debt have been estimated using one of the above explained methods, the overall cost of capital can be calculated, as the weighted average cost of capital.

There are issues related to this stage of the calculation, in particular due to the fact that the debt ratios for private firms are stated in book value terms, rather than market value. Furthermore, the debt ratio for a private firm that plans to go public might change as a consequence of that action.

The most natural assumption to make is that the target leverage of the company will converge to that of the industry sector of pertinence, in a reasonable length of time. This assumption allows to express the debt-to-value ratio as

$$\frac{\tilde{D}}{\tilde{V}} \cong \frac{D_{AVG}}{V}$$

where

$\tilde{D}$ is the debt of the private firm
$\tilde{V}$ is the total value of the private firm
$D_{AVG}$ is the industry average value of debt

Otherwise it is possible to assume that the company will move towards its optimal debt ratio, so that

$$\frac{\tilde{D}}{\tilde{V}} \cong \frac{D_{OPT}}{V}$$

where

$D_{OPT}$ is optimal value of debt

If the analysis assumes that the company will move to the average leverage ratio, the beta must be calculated using the industry average leverage, and estimate the debt rating by using the market average as a proxy.

## 4.3 Growth and Terminal Value

After calculating the present value of the series of cash flows predicted for the first number of years ahead, the valuation must rely on some finalized value that takes care of the infinite sequence of cashflows ahead of that.

The terminal value in the valuation of discounted cash flows represents the estimated value of the potential company's cash flow to be realized after the final discrete projection period.

Such a value resembles the sum of the forecasted future cash flows, which are supposed to grow at a rate that is a permanent growth rate. It follows that the assessment of the terminal value is based on the calculation of the present value of a growing perpetuity.

The basic assumption is that the future cash flows will grow at a constant growth rate over an infinite time. Such an assumption is quite strong and is at the basis of the whole discounted cash flow modelling.

The most commonly used method to calculate the terminal value in the standard DCF model is the Gordon growth model (GGM), which can be described by the equation

$$PV = \frac{NCF_0(1+g)}{r_e - g} \tag{4.2}$$

where

$PV$ is the present value
$NCF_0$ is the net cash flows in the final discrete projection period
$g$ is the selected long-term growth (LTG) rate
$r_e$ is the selected cost of capital

As a first step in terminal value calculation, a growth model should be implemented in order to estimate the permanent growth to be associated with the infinite series of cash flows. It is the normalized long-term series of cash flows (terminal period FCFF or FCFE) during the estimated projection period.

The cash flow stream should reflect the assumption of stable, normalized returns for the company. This is a very important passage in order to ensure a fair valuation of the growth potential of the business.

After that, the terminal cash flow is capitalized and discounted at the appropriate rate, depending on the nature of the cash flow itself. The discounting rate, expressed as the cost of capital indicates the risk inherent.

One important assumption of the Gordon growth model is that the terminal value estimation is based on the assumption of a change in the FCF which is constant over time and never changing itself.

The good proxy for the capitalization rate is the company's Weighted Average Cost of Capital (WACC), which incorporates the risk embedded in the business cash flows, net of the selected long term growth rate (which incorporates the expected growth of the company cash flow).

The capitalization rate calculated as above is then used to discount the projected terminal period's NCF, in order to determine the terminal value, to be added to the discounted previous cash flows (Pratt and Grabowski 2010).

The terminal value is calculated at the end of the first stage, on a cash flow that is following the discrete projection period. It therefore a future value itself and needs to be further discounted to present value.

It is obvious that the final discounting does not take into account the growth rate but simply discounts the cash flow for a discrete amounts of period based on the same discounting rate used for the cash flows in the first stage.

The assumption of a permanent positive or negative growth rate for a business may appear too strong and unrealistic. This is particularly the case when an analyst decides to use a growth rate above the projected inflation rate in the economy.

This implies in fact that the company will grow to infinite, and it will do it on an infinite time span. However, the majority of the terminal value is generated by the cash flow that occurs within the first few periods beyond the discrete projection periods.

The choice of the permanent growth rate for either FCF or Gordon model, implies a long-term view on the cash flows of the company beyond a fairly long period after the terminal period.

The terminal value comes from the accruing spread between the discounting rate and the growth rate at the denominator of the perpetuity calculation, for the whole terminal period following the discrete cash flows period.

That means that for a given WACC or cost of equity discount rate, a lower LTG implies that a higher slice of the terminal value is determined by the near term, while a higher LTG rate causes the opposite result.

A higher LTG in fact corresponds to a steep increase in the cash flow projections in the future. Following this approach, the future periods tend to correspond to higher present values.

For a higher LTG rate the proportion of terminal value generated by cash flows in later periods is much higher, with an implied uncertainty given the distance in time from the present.

There are several issues concerning the right choice of permanent growth then. Such a rate should for example be matched to the inputs considered in the calculation of the cash flows to be discounted.

Another important consideration is the choice of appropriate factors in the choice of rate, and the exclusion of the irrelevant factors from the analysis, in a process that requires careful selection.

Sometimes the analysis behind the choice of the LTG rate mainly relies on factors like historical and projected growth, inflation or any other similar metric that can be directly measured.

It is not possible to account for all factors that affect the growth rate, which makes the estimation of the growth rate an approximate task. Understanding what the factors are is important because the period covered by the LTG rate variable within the Gordon model may start so far into the future that there could be problems.

The projection of the permanent growth rate is very difficult and sometimes it is very hard to identify and select the specific company initiatives that may contribute to the overall growth of the business.

According to Investment Valuation, Tools and Techniques for Determining the Value of Any Asset by Aswath Damodaran, a company's LTG rate should be "determined by a number of subjective factors—the quality of management, the strength of a firm's marketing, its capacity to form partnerships with other firms, and the management's strategic vision, among many others."

One way to estimate the LTG is to look at the firm's existing assets, both tangibles and intangibles. The long-term assets of the company are normally the most addressed factors by analysts when looking for the permanent growth.

Among the assets that analysts normally include in the assessment of the expected growth there are

- The overall industry growth
- The company's market share
- Inflation
- The growth of the existing assets prior to the terminal period.

The growth of the industrial sector in which the company is operating is also an important factor in the growth calculation. Industrial growth rates are normally considered in different magnitudes for discrete projection period and terminal period. In all cases, the latter is the one to be assumed as the factor for permanent growth calculation.

Another important factor is the company market share, with the firm's market penetration being the key parameter to assess whether the impact on permanent growth is high or low.

Existing assets in the company are sometimes the basis for the estimation of LTG rate. But in most cases, and for most companies, this is not enough. Some businesses produce new assets due to innovation. It follows that if the analyst only considers growth from existing assets in the permanent growth rate, he or she may understate the rate.

It is a difficult task to identify the growth generated from existing assets, and separate it from the growth dependent on new assets. The projection of cash flows in this case should be in fact based on a product not developed yet.

There is also another type of growth that can be achieved by the company, which comes through acquisition. Such an operation in fact may affect the value of even the existing assets, both tangibles and intangibles.

The analysis of LTG from the perspective of growth by acquisitions involves establishing how likely the company is to make acquisitions during the terminal period, and what contribution these will have on the projected growth.

An important factor affecting the growth estimation based on acquisitions is the history of the company in that respect, together with the understanding of the acquisition policy of the company.

At an industry level, it is very important to analyze the average level of acquisition activity, by accessing the merger and acquisition databases to analyze the number of transactions that occurred in the subject company industry in the years preceding the valuation date.

As a final relevant factor, the analysis must include a view on the firm's prospective financial ability to make acquisitions. This task is facilitated by the analysis of the past and current financial statements.

Back to terminal value, it is important to stress the fact that such a value constitutes the largest part of the overall present value of cash flows, so that the potential of the company to make acquisitions in perspective constitutes the core of the analysis.

As oppose as the above mentioned factors, there are also others that should be excluded from the LTG analysis. These factors include speculation, theoretical settings, opportunities and trade-offs that maybe existed as of the analysis date but are not relevant in that they are probably not exploited.

After all qualitative factors have been assessed, there is a large part of the analysis that deals with quantitative factors, which are basically hard data to be used for the LTG estimation.

Useful data may include both company-specific information and projected economic growth, both real and nominal, with the range of available types not limited to just these ones.

In some cases there could be a discrepancy between the response of qualitative data and that of quantitative data in terms of the estimation of the LTG rate. The choice in this case is left to the analyst who has to understand the differences and try to reconcile the different values.

The corporate financial statements are a primary source of information, both in the form of historical financial information, and in the form of management-prepared projections about the financials of the business.

Access to historical financial information becomes very important in case the current conditions in which the business is operating are similar to those observed in the past. In this case, the history provides snapshots of the economic results of the business.

As a form of objective information, also the management projections are a reliable source of understanding, being a reflection of the point of view of officers with deep understanding of the economic factors driving the business. They can sometimes give reasonable predictions of the business for a number of years into the future.

One important factor to consider when analyzing the company's historical financial information is the age of the company. This is because a company in the business for a longer time may exhibit a much more stable economic condition. On the other hand, a longer series of time periods, provides a higher amount of data to analyze.

In case the company has constant returns over a significantly long period, the importance of the recent cash flows become very high. The conclusion in this case is that there will be no major changes and the weight of the trend in historical financial information should be considered relevant.

In order to simplify the analysis, consider the example of three hypothetical companies, each corresponding to different impact of firm specific factors on the usability of their historical financial data, for LTG selection.

The first example is a company A, a new start up showing high profits in the early life and good expectations for the future. The volatility of cash flows is also high with a limited diversification of the corporate revenue which comes from only two big customers, with short-term agreements.

The second company, B, is young as well, with a path of increasing cash flows in time, without being dependent on any particular customers for the cash flow pattern to continue in the future.

The last company, C, is a well established old firm with steady growth state and stable cash flows. The growth rate of the company is consistent with that of the overall industry, and the firm is not planning any major changes in the business plan shortly.

For a company like A, since there are no stable factors linking to its productivity, and cash flows are very volatile, the historical data are less meaningful than for other companies. The very unstable customer relationships make this point even stronger. For such a company, it would be more useful to rely on management projections, industry projections and expectations, and other factors.

In case of company B, history can play a role, even if the company is young. This is because the data available for the company seem to be driven by more stable and predictable economic factors.

It is obvious that the valuation of company C will be very benefited by historical data, which have the most use in this case. With a demonstrated past growth and an actual good average reduced growth, it shows to be very interesting for the analysis.

The specific factors of the company at hand will determine what information will carry the most weight and what quantitative methods the analyst may use to provide an indication of a supportable LTG rate.

Further to the cash flow growth analysis is the estimation of the growth of earnings. This can be done by using several methods, and it is important to analyze them in order to acquire a good understanding of the impact on value.

The historical growth in the earnings per share (EPS) is normally a good starting point for the analysis. Both the company's figure and the business sector figures, as measured by other analysts, have to be known.

Ultimately, all growth in earnings can be traced to two fundamentals—how much the firm is investing in new projects, and what returns these projects are making for the firm.

Following Corelli (2016) among the various methods of EPS growth estimation there are arithmetic and geometric averages, and simple regression models. In all cases one must put attention to the period used in estimation and the management of negative earnings, both having a serious impact on the estimation.

The work of the analysts about growth rate estimation is driven by several propositions. First of all, one must take into account that there is far less private information and far more public information in most analyst forecasts than is generally claimed.

Another proposition states that private information mostly comes from the company itself. A third proposition tells us that there is danger when all analysts barely agree with each other on estimations (lemmingitis) and when they agree to little (in which case the information that they have is so noisy as to be useless).

The inputs involved in the estimation of expected long term growth in EPS are reinvestment rate $R_R$ and return on equity $ROE$ which are calculated as

$$r_R = \frac{I_R}{I_C} = R_R$$

where

$r_R$ is the reinvestment rate
$I_R$ are the retained earnings
$I_C$ are the current earnings

and the return on investment $r_I$ is

$$r_I = ROE = \frac{I_N}{E_B}$$

where

$I_N$ is the net income
$E_B$ is the book value of the equity

One must be aware that small changes in ROE translate into large changes in the expected growth rate. Moreover, the lower the current ROE, the greater the effect of changes in ROE will be.

In the long term the company will hardly sustain growth in earnings due to increase in the ROE. After some level, the improvement in ROE will be almost meaningless to the growth in earnings.

If the change is over multiple periods, the second component should be spread out over each period.

How to value the company then? First of all, let's focus on public companies. A public company is assumed to have infinite life, so that its value $P$ can be calculated as the present value of its cash flows over an infinite time span, as

$$P = \sum_{t=1}^{\infty} \frac{C_t}{(1+\tilde{r})^t}$$

where

$\tilde{r}$ is the discount rate applied to the cash flows.

Given that it is impossible to estimate cash flows forever, the common method is to estimate a series of cash flows for a limited growth period, and then add up a terminal value, in order to account for the following infinite time.

$$P_0 = \sum_{t=1}^{T} \frac{C_t}{(1+\tilde{r})^t} + \frac{TV}{(1+\tilde{r})^T}$$

where

$TV$ is the terminal value

The simple form of stable growth equation for estimating the terminal value represents the firm's cash flows as growing at a constant rate forever, so that the present value of those cash flows is

$$P_0 = \frac{C_1}{\tilde{r} - g}$$

where

$g$ is the expected growth rate

The constant growth rate as so calculated is a stable rate that can never be higher than the growth rate of the economy in which the company operates. All companies have an initial growth which can be high for some periods, and will approach the stable growth at a more mature stage of their life.

The overall economy growth in some country of reference resembles the average of the growth rates of the many companies that populate it. It is therefore important to know the general state of the economy in order to guess a stable growth for the firm.

If one assumes that the economy is composed of high growth and stable growth firms, the growth rate of the latter will probably be lower than the growth rate of the economy.

In the assumption that the company is going to disappear in the future, an implicit conclusion is that the growth rate can be negative, so that the terminal value will be lower.

The chosen growth rate should follow the nature of cash flows and discount rates, so that if those are expressed in nominal terms, also the long term growth should be nominal, and expressed in the currency in which the valuation is denominated. One simple proxy for the nominal growth rate of the economy is the risk-free rate.

It is important to underline that the terminal value is not much linked to stable growth, as it is to the excess returns in the stable growth stage. If one assumes that the ROC will equate the cost of capital in stable growth, the growth rate will not impact on the terminal value.

On the other hand, if the assumption is that the firm will earn positive (negative) excess returns in perpetuity, the terminal value will increase (decrease) as the stable growth rate increases.

The first period of discrete cash flows discounting and high growth can be characterized by several behaviors. In particular, there are four different assumptions that can be made.

The first scenario is that there is no high growth and the firm is already in stable growth. The second scenario assumes high growth for a period, after which the growth rate will drop to stable growth.

The third scenario is similar to the second but involves high growth followed by a decline that will gradually lead to stable growth in a three-stage process. Finally, there is the case where each period will be characterized by different growth and margins.

At the same time, assumptions about excess returns must be made as well. In general, the excess returns will be large and positive in the high growth period and decrease as you approach stable growth (the rate of decrease is often titled the fade factor).

There are several factors determining the growth pattern of a company. They range from company size to barriers to entry, passing through current growth rate. First of all, the bigger the firm, the lower the growth rate, due to reaching limits.

Secondly, past growth is not always a reliable indicator of what growth will be in the future. Long term growth will be always lower than current growth. Barriers to entry help companies maintain large market share and therefore enhance higher returns and corresponding higher growth.

There are peculiar characteristics that a firm in stable growth should have. In particular, the risk of the company should be comparable to a stable growth firm in terms of beta and rating.

The beta should be around one, with a rating of BBB or higher on the debt issue by the firm. Also the debt ratio of the firm might increase to reflect the larger and more stable earnings of these firms. Also, the reinvestment rate of the firm should reflect the expected growth rate and the firm's ROC.

Concluding, always be aware that the terminal value will always be a large proportion of the total value. The point reflects the reality that the majority of the returns from holding a stock for a finite period comes from price appreciation.

The proportion of terminal value on the overall value increases as the growth increases. In some cases, the present value of the terminal value is greater than 100% of the current value of the stock.

## Bibliography

Corelli A (2016) Analytical Corporate Finance, Springer

Damodaran A (2002) Investment valuation: tools and techniques for determining the value of any asset, wiley finance editions. Wiley, New York

Koller T, Copeland TE, McKinsey and Company, ProQuest (2000) Valuation workbook: step-by-step exercises and tests to help you master valuation. Hoboken, N.J.: John Wiley and Sons, Incorporated

Pratt SP, Grabowski RJ (2010) Cost of Capital: Applications and Examples, 4th Edition, Wiley

Pratt SP, Reilly R, Schweis RP (2000) Valuing a business: the analysis and appraisal of closely held companies. McGraw-Hill, New York

West TL, Jones JD (1999) Handbook of business valuation. John Wiley, New York

# Chapter 5
# A Real Case: Facebook

**Abstract** After hiding the declining forecasts for the earnings in the coming quarter, the management and underwriters of Facebook decided to move on and price the IPO at the full projected price, therefore resulting in an overvalued IPO, quite uncommon. The financials of Facebook, as per historical value, allowed the underwriters to sell the stocks to the public at the aimed price of $38, a price that was not sustainable in the medium term, therefore declining immediately after public sale. The work done by the students in Sweden clearly shows how it was possible, by just behaving ethically and taking into account all aspects of the valuation, to achieve a fairer valuation in the range of the 20s, exactly the range exhibited by the market in the weeks after the IPO.

**Keywords** Risk · Business model · Underwriters · SEC · Management · Overpriced

The case of the Initial Public Offering of Facebook is one of the milestones in the history and theory of that kind of financial operation, given the peculiarity of the overall process that led to the company valuation and sale on the market.

From a pedagogical point of view, the IPO process showed quite many controversial aspects, that qualify for the perfect application of the theory studied in this book, given the complexity of the operation, and the strangeness of some decisions made at that time.

This chapter therefore represents the logical conclusion of a journey that starts with the basic understanding of the concept and theory of company valuation, and explores the multitude of methods applicable to the task.



A. Corelli, *Inside Company Valuation*, SpringerBriefs in Finance,
DOI 10.1007/978-3-319-53783-2_5

## 5.1 A Strange IPO

The IPO of Facebook, which took place in 2012, is surrounded by shadows of controversy not yet solved The resistance to acquisition attacks and earlier offers to go public suddenly came to an end in 2011, when Zuckemberg decided it was time for the company to trade its shares.

Starting from early 2006, Mark Zuckerberg turned down several acquisition offers, from the $75,000,000 offer from Viacom in 2006, to the valuation given by independent advisors in 2011, at $50 billion.

The last days before the IPO took place were crucial for the assessment of the final Facebook valuation. And it turned out at the end that the selling price was inflated, due to several circumstances to be analysed in this section.

In February 2012 the company filed the necessary documents to the SEC for approval of the IPO operation. In the initial prospectus it was announced that the company had a total of 845 million monthly users with an account, with the website recording a total of 2.7 billions of daily likes and comments.

The company clearly stated in the report to the SEC how a declining process for the membership figures, as well as the income-related assets, was going on in the last semesters.

As for other companies before, the management of Facebook wanted to protect the ownership and control rights of the early investors, by instituting a dual-class stock structure, in 2009.

The consequences of an IPO were the major concern of the company founders, given that in their view, Facebook has been for years a revolutionary company, and the risk was to change its mission in the name of profit. Moreover, one should not forget about the burden of releasing a company's financial details to the public.

But the growth stage of the company at that time required a huge amount of fresh equity capital to be raised, so that the IPO was a mandatory choice. Facebook needed money, and private companies face restrictions on how much stock they can issue for cash.

As from Corelli (2016) "An attempt had been made in 2011, with the help of Goldman Sachs, to raise equity capital by creating a special investment product to sell the private shares of the company to selected Goldman Sachs's customers.

However, the SEC started investigating the plan immediately, putting pressure on the management of Facebook. The founder then decided that going public was the only solution to the capital need.

In late 2011, the *Wall Street Journal* reported that Facebook was anticipating an IPO valuation of $100 billion, nearly four times more than Google's market cap when it went public in 2004".

The company filed the forms to the SEC for starting the IPO process in early February 2012. The form contained everything an investor needs to know before buying shares at an IPO, including basic financial information and the business model.

The paperwork included information on the core business of Facebook, namely online advertising, at that time not expanded yet to the future secondary business lines. In total the company generated a reported \$1 billion from a \$3.7 billion worth of total revenue, at the time of filing.

The report filed to the SEC also listed the investment banks involved in the operation, consisting in a pool of underwriters guided by Morgan Stanley, and including JPMorgan, Chase & Co. and Goldman Sachs, among other smaller 29 banks.

As from Corelli (2016) "The next step in the path to the IPO was, like for many other companies going public, the road show, the promotional tour where executives drum up support for their IPO before large investors.

The first meeting with investors took place in May 2012, just eleven days before the IPO day. The top management of Facebook met the investor at the Sheraton Hotel in midtown Manhattan.

There was a massive presence of journalists, as expected, and a long line of potential investors eager to understand the potential of the IPO. Inside the meeting, Facebook played a video introducing the business model to special clients of its underwriting banks.

That was the time when the road to IPO started being hindered by unwanted news. While the roadshow was taking place, Morgan Stanley found out Facebook had cut revenue projections, resulting in a nearly unprecedented last-minute correction.

The new internal forecasts affected the roadshow, which is usually functioning as a way to boost the expectations on the IPO. In the case of Facebook, that stage was heavily affected by the bad news.

Bad news about the company's new internal forecasts focused on revenue growing slower than expected, due to high tech users flocking to smartphones faster than the company could serve mobile ads.

The last minute cut in revenues by Facebook resulted in an unprecedented downward revision, characterized by poor timing. As a matter of fact, the public disclosure of such information would have turned the IPO down.

For an IPO of that magnitude in fact, even if corrections in the revenue were quite small, statistically, in IPO showbiz statistics run second to momentum, and nothing kills momentum like a poorly timed downward revision.

Obviously something had to be disclosed, but the issue was about what to disclose. The law requires companies to share all information that would likely influence an investor's decision to buy stock, and some of the underwriters were still advising clients based on figures that Facebook later reported to be wrong.

With only a few days left before the IPO, Morgan Stanley and the other banks decided to save the company from being ashamed of their figures, through a procedure that turned out to be quite controversial and harmful to the banks afterwards.

First of all, Facebook would amend its public birth certificate filed to the SEC, the S-1 form, to include information about mobile usage cutting into revenue. Second, the company would disclose more specific information about the downgraded projections.

This amendment of the official form to the SEC, resulted in just changing three pages out of the 170 pages included in the form. The only way to gather information about the ongoing challenge to serve ads to mobile users, was to have knowledge of those three pages".

Other amendments included the declaration that users of the corporate website were growing faster than publicity ads, given the expansion of the services to smartphones users, and also due to the choice of limiting the number of ads per page.

Still from the report, at page 57 an amendment was written stating that the actual growing trend of mobile technology was developing fast, with users shifting their preferences from computers and laptops to smaller and mobile devices. But no financial consequences of such a persistent trend was ever mentioned anywhere in the updated pages.

The issue about the filings was that it was not mentioned anywhere about the projected lower revenue estimates due to the above reasons. Analysts themselves were not capable to figure out the projected decline in revenues given the reported and filed information, making it impossible to predict such a trend.

Facebook was already projecting to trade at high multiples given its earnings figures was slashing its annual projections. The underwriters were not involved into any type of information disclosure themselves.

In financial markets there is a rule about information disclosure. Such a rule does not allow financial analysts to release, publish or email any type of information about the company in the first forty days after the IPO has taken place.

Such a rule protects investors from being hit by the findings of analysts, which may bias their personal judgement about the investment to undertake. On the other hand, this information may then flow privately to institutional clients without a paper trail.

As from Corelli (2016) "It is clear how such a game of telephone calls and private disclosure became so important in the peculiar situation of Facebook. Analysts had the chance to inform only selected investors about the last minute change in the revenue projections.

After the company's surprising eleventh-hour amendment, the unenviable job of explaining Facebook's revisions to the research analysts fell to Cipora Herman, Facebook's vice president of finance.

Bloomberg information system reports on May 11 that the interest of the institutional investors on the IPO was not as strong as expected. Another report from Reuters on the other hand stated that demand was much higher than the planned offer of stocks, with one large unnamed institutional investor calling around syndicate desks trying to get more shares.

The market in those days was therefore characterized by large institutional investors being fully informed about Facebook's last minute troubles, and retail investors being unaware of them.

This is why the retail investors clamoured for Facebook shares while some large investors were planning a massive short-selling, therefore essentially betting against the stock's buyers.

Telephone calls reached the managers of top analysts, from hedge funds and other big investors declaring they would have short-sold the stocks of Facebook, due to the revelations.

But despite the growing consensus among some large investors that Facebook was overpriced, on May 15, three days before Facebook's market debut, the underwriting banks increased the IPO range from $28–$35 to $35–$38, citing heavy demand. A day later, they increased supply to more than 420 million shares.

The new share and price allocation placed Facebook's valuation at the iconic $100 billion mark".

## 5.2 The Firm's Fundamentals

In his book "Analytical Corporate Finance" of 2016, Corelli reports the findings of his students during the solution of a Master level project in company valuation, with Facebook as the target company.

As from the book "students were given the chance to play with the financial statement of Facebook", and analyse the balance sheet (see Table 5.1), income statement (see Table 5.2), cash flow statement (see Table 5.3), and major ratios (see Table 5.4). The following paragraphs summarize the results obtained in terms of fair pricing.

Based on the financials of the company, it is possible to calculate what the fair price of an IPO of Facebook should have been. In particular, it is also possible to grasp the idea of what went wrong in the actual process of the firm going public.

Several valuation methods are available for company valuation, as summarized in Table 5.5.

**Table 5.1** The balance sheet of Facebook™ as of year 2011

| Balance sheet ($M) | 2011 | 2010 | 2009 | 2008 | 2007 |
|---|---|---|---|---|---|
| *Assets* | | | | | |
| Cash | 1512 | 1785 | 633 | 297 | 305 |
| Short-term investments | 2396 | 0 | – | – | – |
| Receivables | 547 | 373 | – | – | – |
| Inventories | 0 | 0 | – | – | – |
| Other current assets | 149 | 88 | – | – | – |
| Total current assets | 4604 | 2246 | 633 | 297 | 305 |
| Net property, plant, equipment | 1475 | 574 | – | – | – |
| Intangibles | 162 | 96 | – | – | – |
| Other long term assets | 90 | 74 | 477 | 208 | 143 |
| Total long-term assets | 1727 | 744 | 477 | 208 | 143 |
| Total assets | 6331 | 2990 | 1110 | 505 | 448 |
| *Liabilities and equity* | | | | | |
| Accounts payable | 63 | 29 | – | – | – |
| Short-term borrowing | 279 | 106 | – | – | – |
| Other short term liabilities | 557 | 254 | – | – | – |
| Total current liabilities | 899 | 389 | 0 | 0 | 0 |
| Long-term borrowings | 398 | 367 | – | – | – |
| Other long term liabilities | 135 | 72 | 241 | 170 | 174 |
| Total long-term liabilities | 533 | 439 | 241 | 170 | 174 |
| Total liabilities | 1432 | 828 | 241 | 170 | 174 |
| Preferred stocks | 615 | 615 | – | – | – |
| Share capital | 2684 | 947 | 868 | 335 | 273 |
| Retained earnings | 1600 | 600 | 0 | 0 | 0 |
| Total equity | 4899 | 2162 | 868 | 335 | 273 |
| Total liabilities and equity | 6331 | 2990 | 1109 | 505 | 447 |

## 5.3 The Solution to the Case Study

As a quick recall of the basic theory behind valuation, please recall that the Free Cash Flow to Firm (FCFF) valuation method is the most suitable when the company is not dividend-paying and the company's FCF's align with company's profitability within a reasonable time horizon.

Given the potentially infinite life of the company, valuation must be based on a 2-stage procedure where there are $N$ future years (usually 5–10) of growth forecast of cash flows, followed by the final (discounted) terminal value.

In the case study the students were recommended to use an approach of the log-descend type, where the revenues for each comparable company have to be logged, and after that the trend of increase of the log price from one year to each other should be recorded (see Table 5.6).

**Table 5.2** The income statement of Facebook™ as of year 2011

| Income statement ($M) | 2011 | 2010 | 2009 | 2008 | 2007 |
|---|---|---|---|---|---|
| Operating revenue | 3711 | 1974 | 777 | 272 | 153 |
| Cost of revenue | (537) | (354) | (145) | (124) | (41) |
| Gross profit | 3174 | 1620 | 632 | 148 | 112 |
| Selling/general/admin expense | (707) | (305) | (205) | (47) | (113) |
| Research and development | (388) | (144) | (87) | (156) | (113) |
| **EBITDA** | **2079** | **1171** | **340** | **−55** | **−114** |
| Depreciation and amortization | (323) | (139) | (78) | 0 | 0 |
| Other income, net | 19 | 2 | (2) | 0 | (10) |
| **Operating income** | **1775** | **1034** | **260** | **(55)** | **(124)** |
| Share of results of associated companies | 0 | 0 | 0 | 0 | 0 |
| **EBIT** | **1775** | **1034** | **260** | **(55)** | **(124)** |
| Interest income (expense) | (42) | (22) | (10) | (1) | (11) |
| **Pretax income** | **1733** | **1012** | **250** | **−56** | **−135** |
| Taxes | (695) | (402) | (25) | 0 | (3) |
| Net income | 1038 | 610 | 225 | (56) | (138) |
| **ROE** | **24.23%** | **39.43%** | **25.92%** | **0%** | **0%** |

**Table 5.3** The statement of cash flows of Facebook™ as of year 2011

| Cash flows ($M) | 2011 | 2010 | 2009 |
|---|---|---|---|
| Net income | 1038 | 610 | 225 |
| Depreciation and amortization | 323 | 139 | 78 |
| Accounts receivable | (174) | (209) | (112) |
| Accounts payable | 102 | 120 | (7) |
| Inventory | 5 | 0 | 0 |
| Other adjustments | 221 | 23 | (60) |
| **Cash from operating activities** | **1515** | **683** | **124** |
| Capital expenditures | (606) | (293) | (33) |
| Acquisitions and other investing activity | (2400) | (22) | (32) |
| **Cash from investing activities** | **(3006)** | **(315)** | **(65)** |
| Dividend paid | 0 | 0 | 0 |
| Sale or purchase of shares | 1030 | 506 | 209 |
| Increase in short-term borrowing | 603 | 115 | 82 |
| Increase in long-term borrowing | (431) | 160 | (48) |
| **Cash from financing activities** | **1202** | **781** | **243** |
| Foreign exchange adjustment | 3 | (3) | 0 |
| **Change in cash and cash equivalents** | **(289)** | **1149** | **302** |
| Changes in working capital | 5 | (70) | (179) |
| **Free cash flows** | **802** | **408** | **101** |

**Table 5.4** Major ratios of Facebook™ as of year 2011

| Indices | 2011 | 2010 | 2009 | 2008 | 2007 |
|---|---|---|---|---|---|
| **Debt/equity** | 0.29 | 0.38 | 0.28 | 0.51 | 0.64 |
| **Debt/assets** | 0.23 | 0.28 | 0.22 | 0.34 | 0.39 |
| **Equity/assets** | 0.77 | 0.72 | 0.78 | 0.66 | 0.61 |

**Table 5.5** Valuation methods for the firm and equity

| Valuation method | | | |
|---|---|---|---|
| Input | Dividend discount | FCF to equity | FCF to firm |
| *Cash flow* | Dividend | FCFE | FCFF |
| *Expected growth* | In dividends | In FCFE and equity income | In FCFF and operating income |
| *Discount rate* | Cost of equity | Cost of equity | Cost of capital |
| *Steady state* | Constant *g* for dividends | Constant g for FCFE | Constant g for FCFF |

**Table 5.6** An example of growth estimation through historical log growth of revenues

| | Google | | | Linkedin | | | Yahoo | | | Avg. Trend | Risk-free |
|---|---|---|---|---|---|---|---|---|---|---|---|
| | Rev. | Log | Trend (%) | Rev. | Log | Trend (%) | Rev. | Log | Trend (%) | | |
| **2010** | 29,321 | 10.29 | 2.13 | 215 | 5.37 | 12.18 | 6324 | 8.75 | –0.24 | 4.69 | 3.56% |
| **2009** | 23,651 | 10.07 | 0.82 | 120 | 4.79 | 37.67 | 6460 | 8.77 | –1.23 | 12.42 | |
| **2008** | 21,796 | 9.99 | 4.83 | 32 | 3.48 | 52.39 | 6969 | 8.88 | 0.38 | 18.53 | |
| **2007** | 16,594 | 9.72 | 2.81 | 10 | 2.28 | 41.84 | 7209 | 8.85 | 0.93 | 15.86 | |
| **2006** | 10,605 | 9.27 | 6.27 | 5 | 1.61 | | 6426 | 8.77 | 2.34 | 4.30 | |
| **2005** | 6139 | 8.72 | 8.12 | | | | 5258 | 8.57 | 4.72 | 6.42 | |
| **2004** | 3189 | 8.07 | 10.66 | | | | 3574 | 8.18 | 10.66 | 10.66 | |
| **2003** | 1466 | 7.29 | 19.79 | | | | 1625 | 7.39 | 7.78 | | |
| **2002** | 440 | 6.09 | 36.47 | | | | 953 | 6.86 | | | |
| **2001** | 86 | 4.46 | | | | | | | | | |
| **2000** | | | | | | | | | | | |

The last trend represents a fair proxy of the beginning of the stable growth period for all companies and it is then wise to take an average of them, so to obtain a risk-free type of rate.

As reported in Corelli (2016), "The trading of Facebook stocks on the regulated exchange became on May 18th 2012, as planned. However, technical problems delayed the start of trading from 11.00 am to half hour later.

In general, during the first day of trading, many technical problems were experienced, and some orders did not go through due to glitches of various kinds. Moreover, investors got confused as to whether or not their orders were successful.

During the first hours of trading the stock price raised to a maximum of $45, but the initial rally of trade was quite unstable. Due to the tendency of the stock price to stay below the IPO price for most of the day, many underwriters bought back shares to support the price.

If it were not for the technical problems mentioned above, the market price would have fallen below the IPO level already on the first trading day. The closing price of the first day ended up being $38.23, only 23 cents above the IPO price and down $3.82 from the opening bell value.

After a disappointing first day, the trading went on and it was supported by a huge figure in terms of volume, standing at 460 million shares exchanged in the first day of trading.

The IPO ended up raising $16 billion, making it the third largest operation of that kind in U.S. history, behind only General Motors and Visa Inc. The stock price corresponded to a market capitalization higher than most U.S. corporations.

At the time, giants like Amazon and Disney ended up being worth less than the freshly public company. Also the wealth of the founder got considerably high, with Zuckerberg's stocks worth $19 billion.

In the subsequent days, Facebook's share turned out to be bullish on nine of the thirteen days after the IPO, with only four bearish days. On May 21st the closing price was as low as $34, and the price dropped to $31 the day after that.

The stock increased modestly in coming days, and Facebook closed its first full week of trading at $31.91. The stock returned to losses for most of its second full week, and had lost over a quarter of its starting value by the end of May.

By June 6 investors had lost $40 billion in capitalization. The third trading week of Facebook ended up in a low $27.10 even worse than a week previous. The company finished its fourth full week with a modest gain, reaching a closing price of $30.01.

The volatility was so high that analysts started delivering controversial predictions of stock price range. Some of them in fact recommended buying in early June, while others recommended selling, in the same period.

The difficult path of Facebook's shares right after the IPO impacted on the stock market as a whole. The high tech sector suffered from a decrease in prices, and the exchanges as a whole suffered from the bearish trend.

The impact of the Facebook IPO was immediate on the market, affecting other technology companies as well. Investment firms faced considerable losses, as a consequence of the technical troubles.

Bloomberg estimated that retail investors may have lost approximately $630 million on Facebook stock since its debut. UBS alone may have lost as much as $350 million.

The Nasdaq stock exchange offered $40 million to investment firms plagued by offering-day computer glitches.

But the investors' losses were too high to be compensated by such an amount, and the other NY exchange, the NYSE commented severely on the decision of increasing the $3 million reimbursement limit usually observed in these cases.

IBM was hired by Nasdaq to take care of the technical issues and overcome them. The IPO impacted both Facebook investors and the company itself. It was said to provide healthy rewards for venture capitalists who finally saw the fruits of their labor.

The wrong IPO definitely lowered the interest of the investors towards Facebook. It was also the case of all the stakeholders, including employees, who changed their view of the investment.

The situation was very risky for the company, also in view of raising further capital for future investments and expenditures. On top of that, the IPO could jeopardize profits for underwriters who face investors skeptical of the technology industry.

From a legal point of view more than 40 lawsuits were filed regarding the Facebook IPO in the month that followed the actual operation, making the situation even worse for the shareholders and the underwriters.

The latter ones, namely Morgan Stanley (MS), JP Morgan (JPM), and Goldman Sachs (GS) all cut their earnings forecasts for the company in the middle of the IPO roadshow.

It is believed that adjustments to earnings estimates were communicated to the underwriters by a Facebook financial officer, who in turn used the information to cash out on their positions while leaving the general public with overpriced shares".

## Bibliography

Corelli A (2016) Analytical corporate finance, 1st edn. Springer

Papier | Fördert gute Waldnutzung
FSC® C083411

Kolibri 360 GmbH
Ferdinand-Jühlke-Straße 7
99095 Erfurt, Deutschland
produktsicherheit@kolibri360.de